HEME BIOLOGY
The Secret Life of Heme in
Regulating Diverse Biological Processes

HEME BIOLOGY
The Secret Life of Heme in
Regulating Diverse Biological Processes

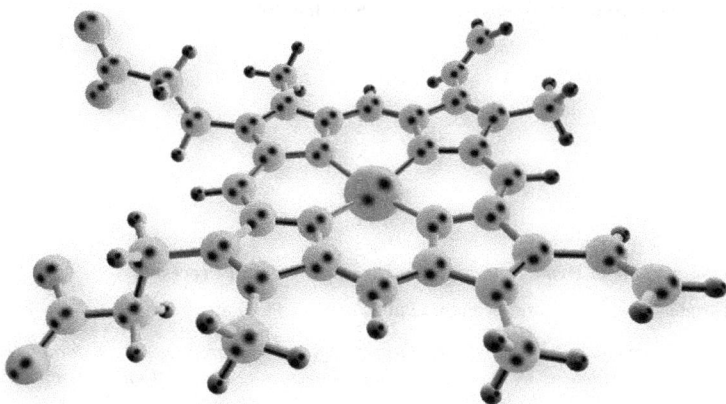

EDITOR

LI ZHANG
The University of Texas at Dallas, USA

W&e World Scientific

NEW JERSEY · LONDON · SINGAPORE · BEIJING · SHANGHAI · HONG KONG · TAIPEI · CHENNAI

Published by

World Scientific Publishing Co. Pte. Ltd.

5 Toh Tuck Link, Singapore 596224

USA office: 27 Warren Street, Suite 401-402, Hackensack, NJ 07601

UK office: 57 Shelton Street, Covent Garden, London WC2H 9HE

British Library Cataloguing-in-Publication Data
A catalogue record for this book is available from the British Library.

HEME BIOLOGY
The Secret Life of Heme in Regulating Diverse Biological Processes

ISBN-13 978-981-4287-92-0
ISBN-10 981-4287-92-X

Typeset by Stallion Press
Email: enquiries@stallionpress.com

Printed in Singapore.

To my mother, Shi-Fong Wu, and father,
Chang-Gui Zhang

PREFACE

Heme biology aims to provide a cohesive understanding of the fundamental regulatory and signaling roles of heme in life and disease processes, as well as the underlying chemical and molecular basis. Heme is arguably the most ubiquitous and unique molecule in the human body. It gives our blood the distinctive color and plays key roles in the proper functioning of all cells. This book provides an in-depth analysis of the roles of heme in three types of human tissues, reticulocytes, brain and liver, on which heme exerts crucial effects.

Heme biology links human function and diseases to the molecular and cellular events occurring in cells, as well as the structural and chemical features of heme. It describes how heme is made, how its level is regulated in various tissues, and what kinds of diseases ensue when the heme level becomes dysregulated. It describes in detail key protein macromolecules with which heme can interact with and thereby control their activities. It also sheds light on how altered interactions between heme and proteins can lead to dysregulated molecular and cellular processes, which in turn cause human diseases. This book includes information about how the chemical properties of heme and heme analogs can be applied to cancer therapy. Overall, this book can be informative and insightful to both the general curious readers and advanced researchers interested in heme biology.

CONTENTS

LIST OF CONTRIBUTORS

Jane-Jane Chen and Rajasekhar NVS Suragani
Harvard-MIT Division of Health Sciences and Technology, Massachusetts Institute of Technology, 77 Massachusetts Avenue, Cambridge, MA 02139

Huiying Ding, Baran D. Sumer, and Jinming Gao
Department of Pharmacology, Department of Otolaryngology, Simmons Comprehensive Cancer Center, University of Texas Southwestern Medical Center, Dallas, Texas 75390

Feng Guo
Department of Biological Chemistry, UCLA School of Medicine, Box 951737, 337 BSRB, Los Angeles, CA 90095-1737

Li Zhang, Yi Li, Rebekah Sessoms, and Donovan Pham Huynh
Department of Molecular and Cell Biology, The University of Texas at Dallas, Mail Stop FO31, 800 W. Campbell Road, Richardson, TX 75080

Aaron Arvey and Christina Leslie
Computational Biology Program, Memorial Sloan-Kettering Cancer Center, 1275 York Ave, Mail Box #460, New York, NY 10065

Tatyana Chernova and Andrew G. Smith
MRC Toxicology Unit, University of Leicester, United Kingdom

Chapter 1

INTRODUCTION

Li Zhang

Overview

Heme, iron protoporphyrin IX (Fig. 1), is arguably one of life's most central molecules. Most of us know about heme because of hemoglobin — the molecule that transports oxygen from the lung to all other organs and tissues in the human body. It is heme that gives hemoglobin the unique oxygen-binding property. The distribution of heme is not even throughout the human body. Roughly 80% of heme in humans is made and present in red blood cells; fifteen percent is made and present in the liver and the rest is distributed in other tissues. All human cells presumably make a basal level of heme for the synthesis and proper functioning of certain proteins and enzymes that use heme as a cofactor or a prosthetic group.

Many living organisms ranging from bacteria to humans can synthesize heme *de novo* (1). Those that do not synthesize heme *de novo*, like *Caenorhabditis elegans*, still require heme for survival and acquire heme via dietary intake (2). Heme and porphyrins have been the subject of fascination and intense studies for scientists for over a century. Porphyrins are compounds composed of a macrocycle of four pyrrole rings linked by four methene bridges. They include metalloporphyrins, such as hemes and chlorophylls. In the early 20th century, scientists started to investigate heme and its porphyrin precursors, due to their association with a class of interesting diseases called porphyrias. Porphyrias are inherited (mostly autosomal dominant) and acquired disorders associated with *partially* defective enzymatic activities of the heme biosynthetic pathway and increased levels of heme precursors (3).

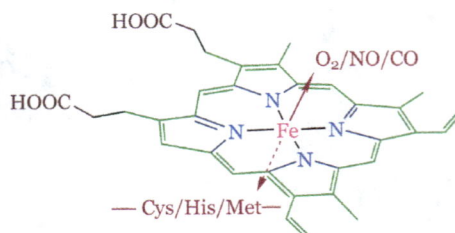

Fig. 1. The structure of heme. Heme is composed of a macrocycle of four pyrrole rings, with four methyl groups, two vinyl groups and two propionate groups attached. The hydrophobic parts of heme are indicated in green. The four nitrogen atoms of pyrrole rings coordinate the heme iron ion. Iron ion can coordinate two axial ligands, which may be Cys, His or Met residue in proteins or small molecules, including oxygen, nitric oxide and carbon monoxide.

It is not confirmed, but is a popular belief that the first documented case of any porphyria disorder dates back to the time of Hippocrates. The term porphyria is derived from the Greek term *porphura* which means "purple pigment" in reference to the color of body fluids in people suffering from a porphyria. Urine that contains porphyrins or porphyrin precursors turns different colors when exposed to the air; black urine became a tell-tale sign of Acute Intermittent Porphyria (AIP). Porphyrias are classified as hepatic or erythropoietic in type, depending on the primary organ in which excess production of porphyrins or precursors takes place (3). Two types of clinical symptoms are associated with porphyrias: cutaneous photosensitivity and acute attacks involving abdominal pains, psychiatric manifestations such as anxiety, depression, and confusion, and neurological manifestations. Erythropoietic porphyrias are associated with only cutaneous photosensitivity; hepatic porphyrias can be associated with both cutaneous photosensitivity and acute attacks. Those associated with acute attacks are also called acute porphyrias.

It has been suggested that persons with congenital erythropoietic porphyrias were the werewolves or vampires of legend. Due to the accumulation of high levels of heme precursors in such subjects, they can have reddish teeth and strong cutaneous photosensitivity. As such, subjects may have skin mutilation, hypertrichosis, and desire to eschew light exposure. This may have led to the superstition of werewolves. Medical records also suggest that many members of the European royal families, including James, IV and I, George III, Frederic the Great of Prussia, and Kaiser

Wilhelm, suffered from acute porphyrias (4). The neurological and psychiatric manifestations associated with acute porphyrias provide a logical explanation for King George's illness. In 1993, Alan Bennett wrote a play, "The Madness of King George" in which he loosely based the King's aliments on porphyria rather than a psychological basis (5).

The latter part of the 20th century has seen a spurt of research on heme and heme proteins due to the importance of heme as a prosthetic group or a cofactor in key proteins and enzymes that support life. These include the following: Hemoglobin and myoglobin that transport and store oxygen; cytochromes and oxidoreductases that support cellular energy generation and biosynthesis; cytochromes P450s that are important for drug metabolism and the synthesis of endogenous substances such as lipids and steroids; cytochrome peroxidases that synthesize molecules critical for innate immune reactions; and oxygenases that synthesize important neuromodulators nitric oxide and carbon monoxide.

More recently, scientists have discovered that heme can serve as a signaling molecule, and thereby regulate a wide array of molecular and cellular processes in living organisms. For example, heme can impact the growth, differentiation and survival of many mammalian cells (6). Heme also controls fundamental molecular and cellular processes, such as protein synthesis, gene transcription, protein localization and assembly. This book is designed to provide you with a complete and up-to-date view of the versatile and fascinating roles of heme in controlling many fundamental biological processes in living organisms, particularly in humans.

To fully understand and appreciate the versatile roles of heme in living processes, it is necessary to have a clear understanding of the structure and chemistry of the heme molecule. Figure 1 illustrates the structure and key chemical features of heme. Heme is composed of a macrocycle of four pyrrole rings. The four nitrogen atoms chelate one iron ion. Iron ion can be in the ferrous (Fe^{+2}) or ferric state (Fe^{+3}). The word heme is usually used as a generic term to identify both ferrous and ferric forms of iron protoporphyrin IX. Properly, however, heme refers only to ferrous protoporphyrin IX, whereas ferric protoporphyrin is called hemin. In air, hemin is more stable than heme. Hemin has a positive charge and is usually isolated with a counterion like chloride.

Heme is a small molecule by molecular mass, and by comparison with macromolecules, such as proteins and nucleic acids in living cells. Yet, it is full of complexity and chemical intricacies. The heme molecule (Fig. 1) contains parts that are highly hydrophobic, including the porphyrin ring and the methyl and vinyl groups, and parts that are hydrophilic, including the iron ion and propionates. Such chemical attributes allow heme to fit in hydrophobic environments as well as hydrophilic ones, and to form hydrophobic interactions as well as salt bridges. Furthermore, the iron ion can adopt and oscillate among several oxidation and electron spin states, enabling heme to freely transfer electrons and to interact with a wide array of molecules, inorganic as well as organic, large and small. Heme is arguably the most chemically and biologically versatile molecule in living organisms.

The properties of the iron ion in heme account for its many biological functions. First, the iron ion can be coordinated by six ligands, but the four pyrroles of heme provide only four ligands. Thus, the heme iron ion can be coordinated by an additional two axial ligands (Fig. 1). This allows heme to associate with proteins and to bind to small molecules such as oxygen, nitric oxide and carbon monoxide. The amino acid residues that chelate the heme iron can be histidine and methionine, as in globins, cytochromes, cytochrome c oxidases and other oxidoreductases. The heme iron can also be coordinated by cysteine, as in cytochrome P450 enzymes. Occasionally, tyrosine can be an axial ligand for the heme iron in proteins. Heme in proteins can be five- or six-coordinated. If the heme iron is five-coordinated in proteins, then the heme iron often binds to small molecules, including oxygen, nitric oxide and carbon monoxide. This allows heme in proteins and enzymes to bind, transport, sense or use these small molecules. The second property of the heme iron is that it can adopt and oscillate among multiple oxidation states, the more stable +2 and +3 states, and the less stable +4 state in certain catalytic intermediates (7). This enables heme enzymes to perform electron transport and oxidation/reduction.

The porphyrin part of the heme molecule also contributes to the biological functions of heme. Four methyl groups, two vinyl groups and two propionate groups are linked to the porphyrin ring in heme (Fig. 1). The methyl groups and vinyl groups, along with the macrocycle, are hydrophobic. Thus, heme is often situated in a hydrophobic pocket in certain enzymes, such as in hemoglobin and c-type cytochromes. The propionate groups often form hydrogen bonds or salt bridges with amino acid residues in enzymes or with solvent molecules. These groups can also allow heme to be covalently attached to proteins and enzymes. For example, in c-type cytochromes, two cysteine residues in the Cys-X-X-Cys-His motif are attached to two vinyl groups in heme. This type of modified heme is referred to as heme C in the literature. The methyl and propionate groups can be modified in various enzymes. Modified forms of heme include heme A and heme D. The detailed structures of these hemes can be found elsewhere (8). The role and structural environment of heme in globins and cytochrome enzymes have been extensively studied and well documented (7).

What is absent, however, from the existing literature is a broad, cohesive, and in-depth analysis of the fascinating roles of heme in diverse biological processes and the underlying molecular bases. This book aims to fill this void. It provides the readers a clear idea about the origin, breadth, and depth of heme biology. In Chapter 2, we explain how heme is synthesized in humans, how its synthesis is regulated in various tissues and organs, and what kinds of diseases can arise when heme biosynthesis becomes defective. In Chapter 3, we describe how heme can control the first step in gene expression, transcription. We show the in-depth molecular mechanisms by which heme controls two master regulators, Hap1 and Bach1. In Chapter 4, we provide a complete view of the role of heme in controlling protein synthesis in red blood cells, and the diseases associated with heme regulation. In Chapter 5, we explain the diverse roles, good and bad, of heme in brain functions. This should allow the readers to appreciate how heme acts to promote neuronal functions and how defective heme function can contribute to various neurological problems in humans. Chapters 6 and 7 describe several recently discovered cases showing the critical roles of heme in regulating key molecules affecting many physiological and diseases processes. Specifically, an essential miRNA processing factor, DiGeorge Critical Region 8 (DGCR8), is regulated by heme. Additionally, two protein tyrosine kinases, Jak2 and Src, are regulated by heme. Defective functioning of these proteins is known to cause major health issues, such as cancer and hematological diseases.

To those readers who want to know more about the chemical and structural characteristics that underlie the diverse roles of heme, Chapter 8 provides an in-depth review of the current literature about heme–protein interactions. Finally, in Chapter 9, we show how the properties of heme biosynthetic pathway and heme precursors can be used in clinical applications. It is amazing that we can take advantage of the photosensitivity of heme precursors to treat serious diseases like cancer, while this property contributes in part to the problems associated with porphyrias. Covering such broad areas associated with heme and also providing in-depth analyses of heme signaling and heme-protein interactions, this book can be informative and useful for both general and expert readers.

References

1. Ponka P. 1999. Cell biology of heme. *Am J Med Sci* 318: 241–256.
2. Rajagopal A, Rao AU, Amigo J, Tian M, Upadhyay SK, Hall C, Uhm S, Mathew MK, Fleming MD, Paw BH, Krause M, Hamza I. 2008. Haem homeostasis is regulated by the conserved and concerted functions of HRG-1 proteins. *Nature* 453: 1127–1131.
3. Anderson KE, Sassa S, Bishop DF, Desnick RJ. 2009. Disorders of heme biosynthesis: X-linked sideroblastic anemia and the porphyrias. In *The Metabolic and Molecular Bases of Inherited Disease*, eds. CR Scriver, AL Beaudt, WS Sly, D Valle, C Barton, KW Kinzler, B Vogelstein, Chapter 124, pp. 1–53. New York: The McGraw-Hill Companies, Inc.
4. Moore MR. 1990. Historical introduction to porphyrins and porphyrias. In *Biosynthesis of Heme and Cholorophylls*, ed. HA Dailey, pp. 1–54. New York: Green Pub. Associates and Wiley-Interscience.
5. Rich F. 1993. Review/Theater: The Madness of King George; Creating a Lovable George III. In *The New York Times*, pp. 1. New York.
6. Mense SM, Zhang L. 2006. Heme: A versatile signaling molecule controlling the activities of diverse regulators ranging from transcription factors to MAP kinases. *Cell Res* 16: 681–692.
7. Messerschmidt A, Huber R, Poulos T, Wieghardt K. 2001. *Handbook of Metalloproteins*. West Sussex: John Wiley & Sons Ltd.
8. Ortiz de Montellano PR. 2009. Hemes in Biology. In *Wiley Encyclopedia of Chemical Biology*, pp. 240–249. West Sussex: John Wiley & Sons Ltd.

Chapter 2

HEME BIOSYNTHESIS AND DEGRADATION

What Happens when it goes Haywire?

Li Zhang and Rebekah Sessoms

Overview

The biosynthesis of heme involves eight enzymatic reactions that convert glycine and succinyl-CoA to heme (1). Heme biosynthesis is not evenly distributed throughout the body. All human cells require heme and are presumed to have the ability to synthesize it. The major sites of heme biosynthesis in the human body are in the bone marrow erythrocytes and liver. The levels of heme in various tissues of the human body fluctuate according to the existing level of heme and the need for additional heme. Intracellular heme levels are precisely and dynamically controlled by modulating the rates of heme biosynthesis and degradation. In this chapter, we describe how heme is synthesized and degraded in humans. We also explain what diseases may ensue if heme biosynthesis or degradation is defective in humans. Such defects may affect adults as well as newborns.

Heme is synthesized by the actions of eight enzymes in humans

In humans, like in yeast, the biosynthetic pathway of heme consists of eight enzymes that progressively convert glycine and succinyl-CoA to heme (Fig. 1) (1–4). In humans, there are two sets of enzymes devoted to heme biosynthesis: housekeeping and erythroid-specific. Housekeeping enzymes are responsible for heme biosynthesis in all but erythroid cells.

7

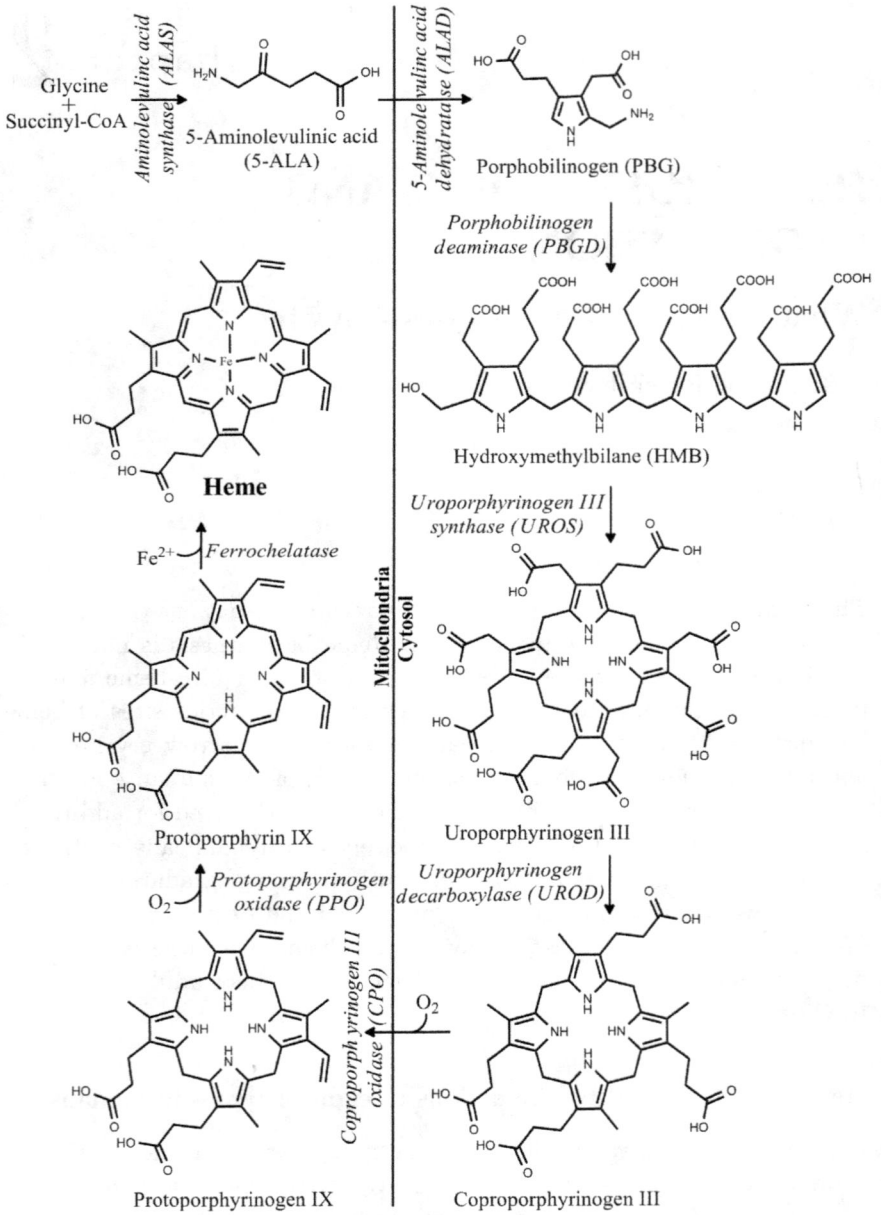

Fig. 1.   The heme biosynthetic pathway in humans.

Erythroid-specific enzymes are responsible for heme biosynthesis only in erythroid cells, which are precursors for red blood cells.

In human cells, the first and the last three steps of heme biosynthesis occur in the mitochondria, and the other four steps occur in the cytosol (1) (Fig. 1). The citric acid cycle (also known as the TCA or Krebs cycle) provides the succinyl CoA needed for heme biosynthesis, and glycine is an amino acid that is readily found in the cell. The citric acid cycle takes place in the mitochondria, making the mitochondria a convenient location for the first step in heme biosynthesis. A condensation reaction between glycine and succinyl-CoA, resulting in 5-aminolevulinic acid (ALA) (Fig. 1), takes place in two steps with a Schiff base intermediate, and is catalyzed by 5-aminolevulinic acid synthase (ALAS). The housekeeping enzyme is designated as ALAS1, while the erythroid-specific enzyme is designated as ALAS2. The genes encoding ALAS1 and ALAS2 are located on chromosomes 3 and X, respectively (5, 6). ALAS1 is expressed in virtually all cells, while ALAS2 is expressed only in fetal liver and adult bone marrow. Their expression is under the control of different sets of promoter elements and regulatory proteins.

The second step in heme biosynthesis is carried out by ALA dehydratase (Fig. 1), which catalyzes the condensation of two molecules of ALA to form porphobilinogen (PBG). In humans, one single gene on chromosome 9 encodes for ALA dehydratase (7). Nonetheless, two different enzymes are expressed in erythroid and non-erythroid cells. This is because the gene contains two promoter regions and two alternative coding exons (8). Both the erythroid-specific and the non-erythorid, housekeeping transcripts contains 12 exons, with 11 of them identical. The first exon in these transcripts is different. Zinc ions serve as cofactors in ALA dehydratase. Lead ions can therefore inhibit ALA dehydratase by displacing the zinc ions (9).

The third enzyme in heme biosynthesis is porphobilinogen deaminase (Fig. 1). It catalyzes the condensation of four molecules of PBG in a head-to-tail manner to generate hydroxymethylbilane (HMB), a linear tetrapyrrole (Fig. 1). Like ALA dehydratase, two forms of porphobilinogen deaminase are encoded by one single gene in humans (10). The gene is located on chromosome 11. Two different transcripts, housekeeping and erythroid-specific, are generated from the gene due to alternate splicing (11). The transcripts are different in their first exons.

The fourth step in heme biosynthesis is the intramolecular rearrangement and ring closure of HMB to form uroporphyrinogen III, a cyclic tetrapyrrole with eight carboxylic side chains (Fig. 1). It is catalyzed by uroporphyrinogen III synthase (UROS). The coding gene for uroporphyrinogen III synthase is located on chromosome 10 (12). Two isoforms of the enyzme are generated in erythroid and non-erythroid cells due to alternate splicing (12). Two different promoters control the expression of these two isoforms.

Uroporphyrinogen decarboxylase (UROD) catalyzes the decarboxylation and the formation of methyl groups (Fig. 1). It enables the formation of coproporphyrinogen III. The gene encoding this enzyme is located on chromosome 1 (13, 14). Unlike the genes encoding the first four heme biosynthetic enzymes, this gene contains only one promoter. The next step in heme biosynthesis is the decarboxylation and oxidation of coproporphyrinogen III, to form protoporphyrinogen IX (Fig. 1). This is catalyzed by coproporphyrinogen oxidase (CPO). The human *CPO* gene is located on chromosome 3, spanning approximately 14 kb with 7 exons (15). Like the gene encoding uroporphyrinogen decarboxylase, the *CPO* gene contains only one single promoter (16).

The seventh enzyme involved in heme biosynthesis is protoporphyrinogen oxidase (PPO) (Fig. 1). It catalyzes the formation of methene bridges in protoporphyrin IX. The human *PPO* gene is mapped to chromosome 1 (17, 18). In the last step of heme biosynthesis, ferrous iron (Fe^{2+}) is inserted into the center of protoporphyrin IX by ferrochelatase, and heme is formed. In humans, there is only one single functional gene, located on chromosome 18 (19). It spans approximately 45 kb and contains 11 exons (20). Notably, two of the enzymes, CPO and PPO, require oxygen as a substrate (Fig. 1). As such, the rate of heme biosynthesis may be controlled by oxygen levels in a certain range. Indeed, in yeast, heme biosynthesis is under the precise control of oxygen levels (4, 21, 22).

Heme biosynthesis is regulated according to the needs for heme in human cells

In the human body, liver and bone marrow erythrocytes are the two major sites of heme biosynthesis. Liver synthesizes 15% of heme while erythrocytes make 80% of heme in a human body (1, 23, 24). In the liver, the majority of heme is used to synthesize cytochrome P450, which carries out biological oxidation (23, 24). Cytochorme P450 detoxifies xenobiotics, such as drugs, lipophilic compounds, environmental contaminants like benzene and toluene, and food additives (25–29). Therefore, in the liver, heme biosynthesis fluctuates according to the demand in detoxifying xenobiotics.

In the liver, heme biosynthesis is controlled at the first step, the synthesis of 5-ALA (Fig. 1), under a variety of conditions (30, 31). In the normal liver, ALAS1 activity is the lowest among all heme biosynthetic enzymes (32). Hence, heme biosynthesis can be controlled by controlling the levels of ALAS1. Heme itself is a crucial regulator of ALAS1 levels. Heme impacts the synthesis of ALAS1 at multiple steps. As illustrated in Fig. 2, ALAS1 synthesis involves several major steps, including transcription, translation and translocation. High heme concentrations inhibit these steps, whereas the lack of heme promotes these steps (33–36). Although heme is synthesized in the mitochondria, all evidence indicates that there exists a "free" intracellular heme pool that mediates the regulation of many processes in diverse cellular compartments (1). How this free heme pool is formed is not yet clear.

In bone marrow erythrocytes, heme is used to synthesize hemoglobin. Hence, in erythrocytes, heme biosynthesis is regulated based on the need for hemoglobin and is subject to erythroid-specific regulatory mechanisms. For example, in erythrocytes, ALAS2, not ALAS1, is responsible for heme biosynthesis. Its expression is regulated by erythroid-specific regulators, such as transcriptional factors like GATA-1 (37). Hence, ALAS2 is not affected by various drugs and chemicals that affect ALAS1. Furthermore, heme does not inhibit ALAS2 synthesis. Rather, heme increases the levels of ALAS2 and other heme biosynthetic enzymes in

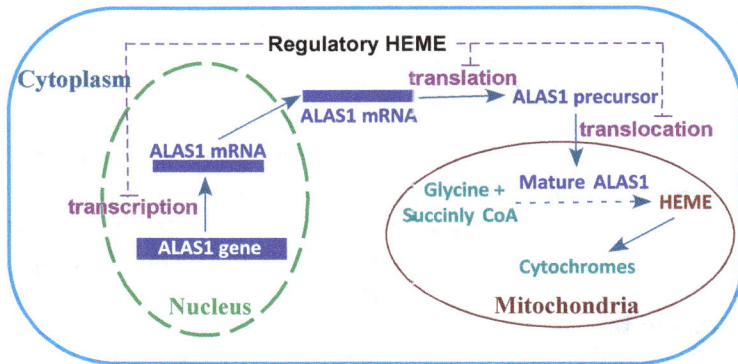

Fig. 2. A cartoon illustrating the regulation of heme biosynthesis in liver cells. In the liver, heme biosynthesis is regulated by the control of the first heme biosynthetic enzymes, ALAS1. The steps involved in the synthesis of ALAS1 is shown. Heme is known to affect three major steps in ALAS1 synthesis.

murine erythroleukemia (MEL) cells. ALAS2 activity in erythrocytes is generally not the rate-limiting step in heme biosynthesis. The increase of heme biosynthesis in erythrocytes is often accompanied by the induction of many heme biosynthetic enzymes (38, 39). When iron is limiting, heme biosynthesis can be limited at the last step, catalyzed by ferrochelatase (Fig. 1) (39). During development, once growth factors such as erythropoietin (40) initiate erythroid differentiation, erythrocytes will make heme continuously for synthesizing hemoglobin. Therefore, heme biosynthesis in erythrocytes is controlled mainly by the availability of iron, and is largely unaffected by external factors.

Defective heme biosynthesis causes serious diseases in humans

Heme is essential for human cell survival; thus defects or mutations that completely abolish heme biosynthesis have never been found. However, even partially defective heme biosynthesis can cause serious problems in humans. A defect in any one of the eight enzymes involved in heme biosynthesis (Fig. 1) causes diseases (1, 41–44). Because iron is required for heme biosynthesis, lack of iron causes iron deficiency anemia. Defects in the erythroid-specific ALAS2 enzyme causes *sideroblastic anemia*, which is a rare, X-linked recessive disease (45, 46).

The clinical manifestations of sideroblastic anemia result from anemia, which is caused by reduced heme biosynthesis and cellular toxicity in various organs due to increased iron levels. Severe symptoms of the anemia include pallor, shortness of breath, fatigue and weakness (46–51). In affected females, symptoms may present a decade later than males, due to protection from iron overload afforded by blood loss during menstruation. Affected patients have serum ferritin levels higher than 200 microgram per deciliter, and transferrin saturation more than 30%. Treatment strategies involve mainly optimizing heme biosynthesis and/or reversing iron overload (49, 51). Heme biosynthesis can be enhanced by pyridoxine and folic acid therapy, while iron overload can be reduced by using phlebotomy and/or chelating therapy.

Defects in the other seven enzymes would lead to *porphyrias*. Defects in the seven enzymes, starting with ALA dehydratase (Fig. 1), cause the following forms of porphyria respectively (1):

1. 5-aminolevulinic acid dehydratase porphyria (ADP), an autosomal recessive hepatic disorder;

2. acute intermittent porphyria (AIP), the most common acute hepatic porphyria and an autosomal dominant disorder;
3. congenital erythropoietic porphyria (CEP), an autosomal recessive disorder;
4. porphyria cutanea tarda (PCT), the most common porphyria;
5. hereditary coproporphyria (HCP), an autosomal dominant hepatic porphyria;
6. variegate porphyria (VP), an autosomal dominant hepatic porphyria; and
7. erythropoietic protoporphyria (EPP), an autosomal dominant disorder.

Patients with porphyrias often accumulate a large amount of heme precursors. Even a 50% reduction in the activity of enzymes downstream of ALAS can cause a significant accumulation of heme precursors. Thus, the effects of these defective enzymes will be twofold. It will cause heme deficiency and the accumulation of heme precursors, both of which cause problems in humans. The causes of porphyrias can be *inherited or acquired*. Inherited porphyrias are those caused by mutations in one of the enzymes, while acquired porphyrias are due to chemicals that affect enzyme activity. For example, lead poisoning is due to its effects on ALA dehydratase and ferrochelatase (52). Porphyrias are also classified as hepatic porphyria and erythropoietic porphyria, based on the sites where the enzymatic defect is located (see Table 1) (1, 41–44).

There are two kinds of symptoms for porphyrias (1, 41–44, 53): photosensitivity and neurological disturbances. Because heme precursors are photosensitive, the accumulation of these compounds in human tissues will cause photosensitivity such as reddish teeth and skin lesions after exposure to sunlight. Another kind of symptom of porphyrias is neurological disturbances associated with acute attacks (1, 41). The neurological manifestations include severe abdominal pain, vomiting, constipation, hypertension, tachycardia, and bladder dysfunction, which have been ascribed to autonomic neuropathy. Additional symptoms include motor weakness and sensory involvement, which correlate with peripheral neuropathy, and mental symptoms without clear morphological findings in the cerebrum.

Erythropoietic porphyrias are associated with only photosensitivity. There are two erythropoietic porphyrias (1, 54, 55): congenital erythropoietic porphyria (CEP), an autosomal recessive disorder caused by the markedly deficient activity of uroporphyrinogen III synthase, and erythropoietic protoporphyria (EPP), an autosomal dominant disorder resulting

Table 1. Characteristics of porphyrias and human heme biosynthetic enzymes.

Porphyria	Deficient enzyme	Gene symbol	Chromosomal location	Inheritance	Main symptoms	Biochemical Findings		
						Erythrocytes	Urine	Stool
Erythropoietic								
Congenital erythropoietic porphyria (CEP)	Uro-porphyrinogen III cosynthase	UROS	10q25.2	Autosomal recessive	Photosensitivity	Uroporphyrin Coproporphyrin	Uroporphyrin Coproporphyrin	Coproporphyrin
Erythropoietic protoporphyria (EPP)	Ferrochelatase	FECH	18q21.3	Autosomal dominant	Photosensitivity	Protoporphyrin	Absent	Protoporphyrin
Hepatic								
ALA dehydratase deficiency porphyria (ADP)	ALA dehydratase	ALAD	9q34	Autosomal recessive	Neurovisceral	Zn-Proto-porphyrin	ALA, Coproporphyrin	
Acute intermittent porphyria (AIP)	PBG deaminase	PBGD	11q23.3	Autosomal dominant	Neurovisceral		ALA, PBG, Uroporphyrin	
Hereditary coproporphyria (HCP)	Copro-porphyrinogen oxidase	CPO	3q12	Autosomal dominant	Neurovisceral photosensitivity		ALA, PBG, Coproporphyrin	
Variegate porphyria (VP)	Proto-porphyrinogen oxidase	PPO	1q23	Autosomal dominant	Neurovisceral and photosensitivity		ALA, PBG, Coproporphyrin	Coproporphyrin Protoporphyrin
Porphyria cutanea tarda (PCT) Isocopropophyrin	Uro-porphyrinogen decarboxylase	UROD	1p34	Variable	Photosensitivity		Uroporphyrin, 7-carboxylate porphyrin	
Hepatoerythropoietic porphyria (HEP) Isocopropophyrin	Uro-porphyrinogen decarboxylase	UROD	1p34	Autosomal recessive	Photosensitivity and neurovisceral	Zn-Proto-porphyrin	Uroporphyrin, 7-carboxylate porphyrin	

from the decreased activity of ferrochelatase (Table 1). Hepatic porphyrias can be associated with both photosensitivity and acute attacks. There are six hepatic porphyrias (Table 1):

1. 5-aminolevulinic acid dehydratase porphyria (ADP), an autosomal recessive hepatic disorder, caused by the deficient activity of 5-aminolevulinic acid dehydratase (Fig. 1);
2. acute intermittent porphyria (AIP), the most common acute porphyria, an autosomal dominant disorder resulting from the half-normal activity of porphobilinogen deaminase;
3. porphyria cutanea tarda (PCT), the most common porphyria, caused by the decreased activity of uroporphyrinogen decarboxylase;
4. hepatoerythropoietic porphyria (HEP), an autosomal recessive disease caused by the marked systemic deficiency of uroporphyrinogen decarboxylase;
5. hereditary coproporphyria (HCP), an autosomal dominant porphyria resulting from the half-normal activity of coproporphyrinogen oxidase;
6. variegate porphyria (VP), an autosomal dominant porphyria due to the half-normal activity of protoporphyrinogen oxidase.

Porphyria cutanea tarda (PCT) is the most common porphyria. It results from decreased uroporphyrinogen decarboxylase activity in the liver. Porphyrins accumulate in large amounts in the liver, and are increased in the plasma and urine (Table 1) (1). Besides mutations in the enzyme, many other factors can reduce the activity of uroporphyrinogen decarboxylase, such as alcohol, hepatitis C infection, estrogen, HIV, smoking, and increased hepatic iron content. Porphyria cutanea tarda is classified into three subtypes: type I (sporadic), type II (familial, autosomal dominant), and type III (familial, rare) (1, 56–58). Type I PCT, known as the "sporadic" form, is the most prevalent form of porphyria worldwide, affecting approximately 1 person out of every 25 000 in the US and 1 in 5000 in Czechoslovakia (1, 56). It can be acquired at any time during the patient's lifetime.

Several precipitating factors may induce PCT, particularly hepatitis C infection (59–62). In southern Europe and in some locations in the United States, 80% or more of patients with PCT are chronically infected with the hepatitis C virus (1). Some patients with PCT also have HIV, suggesting that HIV is also a risk factor (63, 64). Other factors include alcohol usage, oral contraceptives, hormones, hexachlorobenzenes, and increased levels of iron in the blood (53, 65–71). All these factors that increase the need for cytochrome P450 and heme may induce PCT and other porphyrias. Hexacholorobenzene, a fungicide, is known to induce cutaneous porphyria. A large outbreak of cutaneous porphyria occurred in eastern Turkey due to the ingestion of wheat treated with hexacholorobenzene. The occurrence of PCT is higher in males than in females, likely due to higher alcohol intake. Estrogen-containing oral contraceptives and postmenopausal estrogen use are often the cause of PCT in women.

Type II PCT, the "familial" form, results from a genetic mutation in uroporphyrinogen decarboxylase. The homozygous dominant form of this mutation leads to hepatoerythropoietic porphyria (72). It is very rare, and is characterized by the production and buildup of surplus porphyrins not only in hepatocytes, but also in the bone marrow. Type III is the rarest of the types, and the origin of the disorder associated with this type is unknown. Type III is similar to type II, but only has decreased activity in the liver (1).

Symptoms and signs of porphyria cutanea tarda (PCT) include cutaneous manifestations and hepatic abnormalities, but not neurologic involvement. Cutaneous blistering and lesions develop when the skin is exposed to sunlight (1, 68–70). A large number of pus-filled vesicles may form on the outer surfaces of the body, particularly in the front of the hands, on legs, arms, and face, where sun exposure is maximal. The skin becomes extremely fragile and is torn easily, but the healing process is much slower. Oftentimes the symptoms are comparable to those of scleroderma, and misdiagnosis is a possibility. PCT may be characterized further by the growth of excess facial hair, termed hypertrichosis, and too much coloration of the skin, hyperpigmentation (1, 68–70). Skin may thicken and in certain cases, calcify. Increased amounts of ALA, porphobilinogen, and uroporphyrinogen are markedly increased in a PCT patient's urine, plasma, and feces. When these precursors absorb light from the visible portion of the electromagnetic spectrum, they produce reactive oxygen radicals. These species then damage the skin and produce lesions, perhaps through damage to lysosomes and activation of the complement proteins (1, 68–70), which function in lysing and killing cells. PCT is also associated with abnormal liver function tests, such as transaminases and γ-glutamyltranspeptidase. The accumulation of porphyrins in the liver is much higher in PCT than most other porphyrias. PCT also increases the risk of developing hepatocellular carcinoma (73, 74).

All types of PCT (excluding HEP) respond well to treatment by phlebotomy. Red blood cells are removed to deplete iron and to reduce the level of porphyrins in the blood. Administration of chloroquinone or hydroxychloroquine can also be used for treatment in all types of PCT and HEP, if phlebotomies are not appropriate. Lower doses are ideal, approximately 125 mg twice weekly, because a higher dose can cause fever, nausea, and increased synthesis of uroporphyrin (1, 68–70).

Acute intermittent porphyria (AIP) is the most common acute porphyria (1, 75–77). It results from the half-normal activity of porphobilinogen deaminase. It often occurs in women between puberty and menopause. It is precipitated by factors that increase the need for heme biosynthesis, such as sex steroids, contraceptives and other prescribed drugs, alcohol, stress, and fasting. Porphyrias may account for about 10% of pain women experience prior to their menstrual cycle. The United States experiences approximately 1–5 cases of AIP per 100 000 people, while Sweden ranks the highest in AIP patients with an average of 60–100 cases per 100 000 (1, 77, 78).

The clinical manifestations of AIP involve the disturbances of visceral, peripheral, and central nervous systems. Patients with AIP experience acute episodes of neurovisceral problems accompanied by increased excretion of porphyrin precursors (1, 77, 79). Acute attacks often start with severe colicky abdominal pain, vomiting, and constipation (80). Abdominal pain occurs in 85%–95% of acute cases and is the most common sign (81–83). It is usually severe, steady and poorly localized. Tachycardia occurs in up to 80% of acute attacks (83). Peripheral neuropathy involve primarily motor. Muscle weakness often begins in the proximal muscles, and occurs more frequently in the arms than in the legs. Seizures may occur in certain patients experiencing acute attacks (84). Disturbances involving the central nervous system is common but highly variable. Patients can experience severe anxiety, insomnia, depression, disorientation, hallucinations, and paranoia. Suicide is reported to be common.

AIP diagnosis can be diagnosed through a urine sample (1). When compared to normal laboratory values, urine from AIP patients should show increased levels of the porphyrin precursors ALA and PBG (Table 1). Due to the deficient activity of PBGD, only partial amounts of these metabolites can go further in the pathway to synthesize heme. Normal lab values of PBG are approximately 5.3–9.2 nmol/L for women, and 3.4–9.5 nmol/L for men (80, 85). For AIP patients, these values are usually more than doubled. The porphyrins turn the urine dark red in color, and when exposed to ultraviolet radiation from sunlight, the color may deepen to dark purple.

Treatments for acute attacks include infusion of dextrose/glucose and administration of synthetic heme arginate (79, 80, 86). The latter medication is manufactured by only one company in the US and is known by its brand name, Panhematin. Drugs that increase the synthesis of cytochrome P450 and the need for heme, such as barbiturates, sulfonamides, and diphenylhydantoin may precipitate acute attacks and should be avoided (76, 79, 80, 86). Many commonly used drugs such as antibiotics and pain killers can also exacerbate the problems of AIP and should be avoided. An extensive list of drugs that are safe or unsafe in the cute porphyrias can be found at the American Porphyria Foundation's website.

It is also worth noting that a general reduction of heme biosynthesis can be associated with serious human diseases. For example, heme deficiency alters the functions of diverse neuronal cells, and causes defects like those observed in neurodegenerative diseases in neuronal cells (87–89). A recent study shows that the levels of the two rate-limiting enzymes in the heme biosynthetic pathway, ALA synthase and porphobilinogen deaminase (Fig. 1), are significantly reduced in the brains of Alzheimer's disease patients, compared to normal brains (90).

Heme is degraded continuously in the human body

In the normal human body, red blood cells are continuously made and broken down when they reach the end of their life span at around 120 days. Senescent or old red cells are removed and engulfed by the reticuloendothelial system at extravascular sites. Heme is then released from the globin chains and degraded. The first step in heme degradation involves heme oxygenase (Fig. 2) (91, 92). Heme oxygenase selectively cleaves the α-methene bridge of the porphyrin ring and generates biliverdin (blue-green) in the presence of oxygen and NADPH (93). This involves an electrophilic attack of ferrous iron in heme (Fig. 2). Subsequently, the heme ring is opened nonenzymatically by the reaction of oxygen (94). Heme oxygenase activity is highest in the spleen, which disposes senescent erythrocytes (95). In the liver, Kupffer cells and hepatocytes have heme oxygenase activity.

Two heme oxygenase (HO) enzymes, HO-1 and HO-2, have been identified, confirmed and characterized in mammals (96–98). HO-1 expression is inducible, while HO-2 expression is constitutive (99). The expression of each gene differs by cell type, tissue distribution and regulation (100–102). The regulation of HO-1 expression in mammalian cells is very complex. HO-1 is induced by many factors, such as hypoxia, hydrogen peroxide, heavy metals, heme and depletion of cellular glutathione (99, 103, 104). The regulation of HO-1 expression by these factors is not only cell type-dependent but also species-dependent. For example, hypoxia induces HO-1 expression in rodent, bovine and monkey cells, but represses HO-1 expression in several human cell lines, including lung cancer A549 cells, umbilical vein endothelial cells and glioblastoma cells (105–111). Protoporphyrins containing other metals, such as tin, can bind to heme oxygenase with high affinity, but cannot be degraded. They are potent inhibitors of heme oxygenase.

Biliverdin is water soluble and is readily excreted by the liver in many amphibian, avian, and fish species. It is the major bile pigment in these animals. In mammals, however, biliverdin is further reduced to the orange bile pigment, bilirubin, by the enzyme biliverdin reductase. Following its formation in the reticuloendothelial system, bilirubin is released into the blood, and is bound by serum albumin. Subsequently, bilirubin is cleared rapidly by the liver. Bilirubin is a lipophilic, water-insoluble compound (112). It is responsible for the yellow associated with bruises, urine, and the yellow discoloration in jaundiced patients. Heme is the exclusive source of bilirubin. In humans, 250 to 400 mg of bilirubin is formed due to the degradation of heme-containing molecules or free heme (113). About 80% of bilirubin results from the degradation of hemoglobin in senescent erythrocytes (114).

To permit its secretion into bile, bilirubin is rendered water-soluble by conjugation with glucuronic acid (Fig. 3). This reaction is catalyzed by the enzyme bilirubin UDP-glucuronyl transferase (Fig. 3) (115, 116). Heme degradation has several purposes (112, 117, 118). First, it releases iron from heme in hemoproteins and allows iron to be reused. Second, it disposes the porphyrin compounds. Third, it generates carbon monoxide, which is a neuromodulator. Finally, the heme degradation product, bilirubin, is a potent antioxidant (119–121). It can provide protection against oxidative damage, when its level is not high enough to cause toxicity (117, 118). Bilirubin production reflects turnovers of biologically important hemoproteins, particularly hemoglobin. At a steady-state condition of blood hemoglobin levels, the rates of bilirubin production and heme biosynthesis should be about the same.

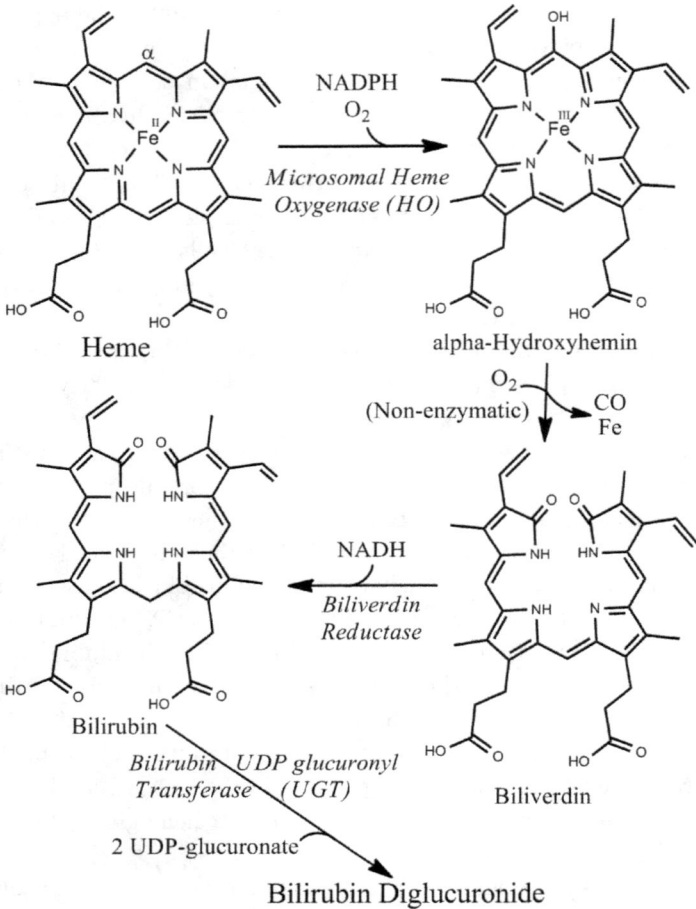

Fig. 3.   The heme degradation pathway in humans.

Defects in heme degradation can cause serious diseases

The most common abnormality associated with defective heme degradation is hyperbilirubinemia, which would result in deposition of bilirubin in the skin, thereby causing jaundice (92). Hyperbilirubinemia can result from increased hemolysis (premature destruction of erythrocytes), such as those occurring in sickle cell anemia, hereditary spherocytosis, and toxic or idiosyncratic drug interactions. Hyperbilirubinemia can also result from malfunction of the liver, for example, damage resulting from hepatitis infection or liver tumors. Another common form of jaundice is neonatal jaundice (122). About half of all neonates may become clinically jaundiced during the first 5 days of their life, largely because bilirubin UDP-glucuronyl transferase activity in the newborn is very low.

To treat neonatal jaundice, phototherapy is used (123). The infants are placed under blue light (400–500 nm) so that bilirubin can be degraded into non-toxic compounds. To treat jaundice in adults, Tin protoporphyrin, which is a potent inhibitor of heme oxygenase, can be used. Without proper management, profound unconjugated hyperbilirubinemia can cause bilirubin encephalopathy (kernicterus) in patients, particularly in neonates (124). Kernicterus usually presents between the 3rd and 6th days of life (125). The signs include poor feeding and feeble suck reflex, high-pitched cry, hyper- or hypotonia, and convulsion. If untreated, this can progress to lethargy, atonia, and death. Long-term adverse effects include asymmetric spasticity, sensorineural hearing loss, delay in motor development, chorioathetosis, paralysis of upward gaze, dental dysplasia, cognitive dysfunction, and mental retardation.

There are several forms of hereditary hyperbilirubinemia (92, 126). They can be classified into two types: predominantly unconjugated and conjugated hyperbilirubinemia (127, 128). The disorders associated with predominantly unconjugated hyperbilirubinemia include Crigler-Najjar syndrome types I and II and Gilbert syndrome (116, 129, 130). These disorders are caused by mutations that cause the synthesis of a defective UGT enzyme or a lower level of this enzyme (Fig. 3) (116, 131, 132). The disorders associated with predominantly conjugated hyperbilirubinemia include Dubin-Johnson syndrome, Rotor syndrome, and benign recurrent intrahepatic cholestasis (133, 134). These disorders are caused by mutations of genes encoding functions required for the secretion of conjugated bilirubin into bile. For example, Dubin-Johnson syndrome is caused by a mutation of bile canalicular multispecific organic anion transporter, involved in the excretion of many non-bile salt organic anions (92).

In summary, heme is a metallo-organic molecule with unique and complex chemical and biochemical properties. As such, complex enzymatic pathways and networks are necessary to maintain its homeostasis in living organisms, particularly in humans. Any perturbation of such enzymatic pathways and networks can cause serious disorders in humans. The next few chapters will further explore the diverse roles that heme plays in controlling fundamental processes underlying health and diseases.

References

1. Anderson KE, Sassa S, Bishop DF, Desnick RJ. 2009. Disorders of heme biosynthesis: X-linked sideroblastic anemia and the porphyrias. In *The metabolic and Molecular Bases of Inherited Disease*, eds. CR Scriver, AL Beaudt, WS Sly, D Valle, C Barton, KW Kinzler, B Vogelstein, Chapter 124, pp. 1–53. New York: The McGraw-Hill Companies, Inc.

2. Ortiz de Montellano PR. 2009. Hemes in Biology. In *Wiley Encyclopedia of Chemical Biology*, pp. 240–249. West Sussex: John Wiley & Sons, Ltd.

3. Sassa S, Kappas A. 1995. Disorders of heme production and catabolism. In *Blood: Principles and Practice of Hematology*, eds. RI Handin, SE Lux, TP Stossel, pp. 1473–1523. Philadephia: J.B. Lippincott Company.

4. Labbe-Bois R, Labbe P. 1990. Tetrapyrrole and heme biosynthesis in the yeast Sacchromyces cerevisiae. In *Biosynthesis of Heme and Cholorophylls*, ed. HA Dailey, pp. 235–285. New York: Green Pub. Associates and Wiley-Interscience.

5. Bishop DF. 1990. Two different genes encode delta-aminolevulinate synthase in humans: Nucleotide sequences of cDNAs for the housekeeping and erythroid genes. *Nucleic Acids Res* 18: 7187–7188.

6. Bishop DF, Henderson AS, Astrin KH. 1990. Human delta-aminolevulinate synthase: Assignment of the housekeeping gene to 3p21 and the erythroid-specific gene to the X chromosome. *Genomics* 7: 207–214.

7. Potluri VR, Astrin KH, Wetmur JG, Bishop DF, Desnick RJ. 1987. Human delta-aminolevulinate dehydratase: Chromosomal localization to 9q34 by in situ hybridization. *Hum Genet* 76: 236–239.

8. Kaya AH, Plewinska M, Wong DM, Desnick RJ, Wetmur JG. 1994. Human delta-aminolevulinate dehydratase (ALAD) gene: Structure and alternative splicing of the erythroid and housekeeping mRNAs. *Genomics* 19: 242–248.

9. Sassa S, Granick S, Bickers DR, Levere RD, Kappas A. 1973. Studies on the inheritance of human erythrocyte delta-aminolevulinate dehydratase and uroporphyrinogen synthetase. *Enzyme* 16: 326–333.

10. Meisler M, Wanner L, Eddy RE, Shows TB. 1980. The UPS locus encoding uroporphyrinogen I synthase is located on human chromosome 11. *Biochem Biophys Res Commun* 95: 170–176.

11. Chretien S, Dubart A, Beaupain D, Raich N, Grandchamp B, Rosa J, Goossens M, Romeo PH. 1988. Alternative transcription and splicing of the human porphobilinogen deaminase gene result either in tissue-specific or in housekeeping expression. *Proc Natl Acad Sci U S A* 85: 6–10.

12. Xu W, Kozak CA, Desnick RJ. 1995. Uroporphyrinogen-III synthase: molecular cloning, nucleotide sequence, expression of a mouse full-length cDNA, and its localization on mouse chromosome 7. *Genomics* 26: 556–562.

13. de Verneuil H, Grandchamp B, Foubert C, Weil D, N'Guyen VC, Gross MS, Sassa S, Nordmann Y. 1984. Assignment of the gene for uroporphyrinogen decarboxylase to human chromosome 1 by somatic cell hybridization and specific enzyme immunoassay. *Hum Genet* 66: 202–205.

14. Dubart A, Mattei MG, Raich N, Beaupain D, Romeo PH, Mattei JF, Goossens M. 1986. Assignment of human uroporphyrinogen decarboxylase (URO-D) to the p34 band of chromosome 1. *Hum Genet* 73: 277–279.

15. Cacheux V, Martasek P, Fougerousse F, Delfau MH, Druart L, Tachdjian G, Grandchamp B. 1994. Localization of the human coproporphyrinogen oxidase gene to chromosome band 3q12. *Hum Genet* 94: 557–559.

16. Takahashi S, Taketani S, Akasaka JE, Kobayashi A, Hayashi N, Yamamoto M, Nagai T. 1998. Differential regulation of coproporphyrinogen oxidase gene between erythroid and nonerythroid cells. *Blood* 92: 3436–3444.

17. Taketani S, Inazawa J, Abe T, Furukawa T, Kohno H, Tokunaga R, Nishimura K, Inokuchi H. 1995. The human protoporphyrinogen oxidase gene (PPOX): Organization and location to chromosome 1. *Genomics* 29: 698–703.

18. Roberts AG, Whatley SD, Daniels C, Holmans P, Fenton I, Owen MJ, Thompson P, Long C, Elder GH. 1995. Partial characterization and assignment of the gene for protoporphyrinogen oxidase and variegate porphyria to human chromosome 1q23. *Hum Mol Genet* 4: 2387–2390.

19. Whitcombe DM, Carter NP, Albertson DG, Smith SJ, Rhodes DA, Cox TM. 1991. Assignment of the human ferrochelatase gene (FECH) and a locus for protoporphyria to chromosome 18q22. *Genomics* 11: 1152–1154.

20. Taketani S, Inazawa J, Nakahashi Y, Abe T, Tokunaga R. 1992. Structure of the human ferrochelatase gene. Exon/intron gene organization and location of the gene to chromosome 18. *Eur J Biochem* 205: 217–222.

21. Mattoon J, Lancashire W, Sanders H, Carvajal E, Malamud D, Braz G, Panek A. 1979. Oxygen and catabolite regulation of hemoprotein biosynthesis in the yeast *Saccharomyces cerevisiae*. In *Biosynthesis of Heme and Cholorophylls*, ed. WJ Caughey, pp. 421–435. New York: Academic Press.

22. Hon T, Dodd A, Dirmeier R, Gorman N, Sinclair PR, Zhang L, Poyton RO. 2003. A mechanism of oxygen sensing in yeast: Multiple oxygen-responsive steps in the heme biosynthetic pathway affect Hap1 activity. *J Biol Chem* 278: 50771–50780.

23. Granick S, Sassa S. 1971. δ-Aminolevulinic acid synthetase and the control of heme and chlorophyll synthesis. In *Metabolic Regulation*, ed. JH Vogel, pp. 77. New York: Academic Press.

24. Sassa S, Kappas A. 1981. Genetic, metabolic, and biochemical aspects of the porphyrias. In *Adv Hum Genet*, pp. 121–231.

25. De Matteis F, Abbritti G, Gibbs AH. 1973. Decreased liver activity of por-phyrin-metal chelatase in hepatic porphyria caused by 3,5-diethoxycarbonyl-1,4-dihydrocollidine. Studies in rats and mice. *Biochem J* 134: 717–727.

26. De Matteis F, Gibbs A. 1972. Stimulation of liver 5-aminolaevulinate syn-thetase by drugs and its relevance to drug-induced accumulation of cytochrome P-450. Studies with phenylbutazone and 3,5-diethoxycarbonyl-1,4-dihydrocollidine. *Biochem J* 126: 1149–1160.

27. Whiting MJ, Granick S. 1976. Delta-Aminolevulinic acid synthase from chick embryo liver mitochondria. I. Purification and some properties. *J Biol Chem* 251: 1340–1346.

28. Sinclair PR, Granick S. 1975. Heme control on the synthesis of delta-aminolevulinic acid synthetase in cultured chick embryo liver cells. *Ann N Y Acad Sci* 244: 509–520.

29. Sassa S, Kappas A. 1977. Induction of aminolevulinate synthase and por-phyrins in cultured liver cells maintained in chemically defined medium. Permissive effects of hormones on induction process. *J Biol Chem* 252: 2428–2436.

30. Granick S. 1966. The induction in vitro of the synthesis of delta-aminole-vulinic acid synthetase in chemical porphyria: A response to certain drugs, sex hormones, and foreign chemicals. *J Biol Chem* 241: 1359–1375.

31. Granick S, Urata G. 1963. Increase in activity of alpha-aminolevulinic acid synthetase in liver mitochondria induced by feeding of 3,5-dicarbethoxy-1,4-dihydrocollidine. *J Biol Chem* 238: 821–827.

32. Hutton JJ, Gross SR. 1970. Chemical induction of hepatic porphyria in inbred strains of mice. *Arch Biochem Biophys* 141: 284–292.

33. Sassa S, Granick S. 1970. Induction of 5-aminolevulinic acid synthetase in chick embryo liver cells in cluture. *Proc Natl Acad Sci USA* 67: 517–522.

34. Tyrrell DL, Marks GS. 1972. Drug-induced porphyrin biosynthesis. V. Effect of protohemin on the transcriptional and post-transcriptional phases of aminole-vulinic acid synthetase induction. *Biochem Pharmacol* 21: 2077–2093.

35. Lathrop JT, Timko MP. 1993. Regulation by heme of mitochondrial protein transport through a conserved amino acid motif. *Science* 259: 522–525.

36. Munakata H, Sun JY, Yoshida K, Nakatani T, Honda E, Hayakawa S, Furuyama K, Hayashi N. 2004. Role of the heme regulatory motif in the

heme-mediated inhibition of mitochondrial import of 5-aminolevulinate synthase. *J Biochem* 136: 233–238.

37. Surinya KH, Cox TC, May BK. 1997. Transcriptional regulation of the human erythroid 5-aminolevulinate synthase gene. Identification of promoter elements and role of regulatory proteins. *J Biol Chem* 272: 26585–26594.

38. Granick JL, Sassa S. 1978. Hemin control of heme biosynthesis in mouse Friend virus-transformed erythroleukemia cells in culture. *J Biol Chem* 253: 5402–5406.

39. Rutherford T, Thompson GG, Moore MR. 1979. Heme biosynthesis in Friend erythroleukemia cells: Control by ferrochelatase. *Proc Natl Acad Sci U S A* 76: 833–836.

40. Fisher JW. 1997. Erythropoietin: Physiologic and pharmacologic aspects. *Proc Soc Exp Biol Med* 216: 358–369.

41. Meyer UA, Schuurmans MM, Lindberg RL. 1998. Acute porphyrias: Pathogenesis of neurological manifestations. *Semin Liver Dis* 18: 43–52.

42. Moore MR. 1990. Historical introduction to porphyrins and porphyrias. In *Biosynthesis of Heme and Cholorophylls*, ed. HA Dailey, pp. 1–54. New York: Green Pub. Associates and Wiley-Interscience.

43. Nordmann Y, Deybach J-C. 1990. Human hereditary porphyrias. In *Biosynthesis of Heme and Cholorophylls*, ed. HA Dailey, pp. 491–541. New York: Green Pub. Associates and Wiley-Interscience.

44. Kappas A, Sassa S, Galbraith RA, Nordmann Y. 1995. The porphyrias. In *The Metabolic and Molecular Bases of Inherited Disease*, ed. CR Scriver, AL Beaudt, WS Sly, D Valle, pp. 2103–2160. New York: The McGraw-Hill Companies, Inc.

45. May A, Bishop DF. 1998. The molecular biology and pyridoxine responsiveness of X-linked sideroblastic anaemia. *Haematologica* 83: 56–70.

46. Cotter PD, May A, Li L, Al-Sabah AI, Fitzsimons EJ, Cazzola M, Bishop DF. 1999. Four new mutations in the erythroid-specific 5-aminolevulinate synthase (ALAS2) gene causing X-linked sideroblastic anemia: Increased pyridoxine responsiveness after removal of iron overload by phlebotomy and coinheritance of hereditary hemochromatosis. *Blood* 93: 1757–1769.

47. Cotter PD, May A, Fitzsimons EJ, Houston T, Woodcock BE, al-Sabah AI, Wong L, Bishop DF. 1995. Late-onset X-linked sideroblastic anemia. Missense mutations in the erythroid delta-aminolevulinate synthase (ALAS2) gene in two pyridoxine-responsive patients initially diagnosed with acquired refractory anemia and ringed sideroblasts. *J Clin Invest* 96: 2090–2096.

48. Cotter PD, Baumann M, Bishop DF. 1992. Enzymatic defect in "X-linked" sideroblastic anemia: molecular evidence for erythroid delta-aminolevulinate synthase deficiency. *Proc Natl Acad Sci U S A* 89: 4028–4032.

49. Cotter PD, Rucknagel DL, Bishop DF. 1994. X-linked sideroblastic anemia: Identification of the mutation in the erythroid-specific delta-aminolevulinate synthase gene (ALAS2) in the original family described by Cooley. *Blood* 84: 3915–3924.

50. Prades E, Chambon C, Dailey TA, Dailey HA, Briere J, Grandchamp B. 1995. A new mutation of the ALAS2 gene in a large family with X-linked sideroblastic anemia. *Hum Genet* 95: 424–428.

51. Edgar AJ, Losowsky MS, Noble JS, Wickramasinghe SN. 1997. Identification of an arginine452 to histidine substitution in the erythroid 5-aminolaevulinate synthetase gene in a large pedigree with X-linked hereditary sideroblastic anaemia. *Eur J Haematol* 58: 1–4.

52. Shoolingin-Jordan PM, Spencer PMS, Erskine PE, Cheung KM, Cooper JB, Norton EB. 2002. 5-Aminolaevulinic acid dehydratase: Metals, mutants and mechanism. *Biochem Soc Trans* 30: 584–590.

53. Poh-Fitzpatrick MB. 2000. Porphyrias: Photosensitivity and phototherapy. *Methods Enzymol* 319: 485–493.

54. Desnick RJ, Glass IA, Xu W, Solis C, Astrin KH. 1998. Molecular genetics of congenital erythropoietic porphyria. *Semin Liver Dis* 18: 77–84.

55. Poh-Fitzpatrick MB, Wang X, Anderson KE, Bloomer JR, Bolwell B, Lichtin AE. 2002. Erythropoietic protoporphyria: Altered phenotype after bone marrow transplantation for myelogenous leukemia in a patient heteroallelic for ferrochelatase gene mutations. *J Am Acad Dermatol* 46: 861–866.

56. Elder GH. 1998. Porphyria cutanea tarda. *Semin Liver Dis* 18: 67–75.

57. Bulat V, Lugovic L, Situm M, Buljan M, Bradic L. 2007. Porphyria cutanea tarda as the most common porphyria. *Acta Dermatovenerol Croat* 15: 254–263.

58. Lambrecht RW, Thapar M, Bonkovsky HL. 2007. Genetic aspects of porphyria cutanea tarda. *Semin Liver Dis* 27: 99–108.

59. Fargion S, Piperno A, Cappellini MD, Sampietro M, Fracanzani AL, Romano R, Caldarelli R, Marcelli R, Vecchi L, Fiorelli G. 1992. Hepatitis C virus and porphyria cutanea tarda: Evidence of a strong association. *Hepatology* 16: 1322–1326.

60. Herrero C, Vicente A, Bruguera M, Ercilla MG, Barrera JM, Vidal J, Teres J, Mascaro JM, Rodes J. 1993. Is hepatitis C virus infection a trigger of porphyria cutanea tarda? *Lancet* 341: 788–789.

61. Chuang TY, Brashear R, Lewis C. 1999. Porphyria cutanea tarda and hepatitis C virus: A case-control study and meta-analysis of the literature. *J Am Acad Dermatol* 41: 31–36.

62. Egger NG, Goeger DE, Miskovsky EP, Weinman SA, Payne DA, Anderson KE. 1999. Multiple risk factors including HFE mutations in porphyria cutanea tarda. *Gastroenterology* 116: Starts at A1206.

63. Wissel PS, Sordillo P, Anderson KE, Kappas A, Sassa S, Savillo RL. 1987. Porphyria cutanea tarda associated with the acquired immune deficiency syndrome. *Am J Hematol* 25: 107–113.

64. Cohen PR. 1991. Porphyria cutanea tarda in human immunodeficiency virus-seropositive men: Case report and literature review. *J Acquir Immune Defic Syndr.* 4: 1112–1127.

65. Taljaard JJ, Shanley BC, Deppe WM, Joubert SM. 1972. Prophyrin metabolism in experimental hepatic siderosis in the rat. II. Combined effect of iron overload and hexachlorobenzene. *Br J Haematol* 23: 513–519.

66. Taljaard JJ, Shanley BC, Deppe WM, Joubert SM. 1972. Porphyrin metabolism in experimental hepatic siderosis in the rat. III. Effect of iron overload and hexachlorobenzene on liver haem biosynthesis. *Br J Haematol* 23: 587–593.

67. Moore AW, 3rd, Coke JM. 2000. Acute porphyric disorders. *Oral Surg Oral Med Oral Pathol Oral Radiol Endod* 9C: 257–262.

68. Sams H, Kiripolsky MG, Bhat L, Stricklin GP. 2004. Porphyria cutanea tarda, hepatitis C, alcoholism, and hemochromatosis: A case report and review of the literature. *Cutis* 73: 188–190.

69. Kostler E, Wollina U. 2005. Therapy of porphyria cutanea tarda. *Expert Opin Pharmacother* 6: 377–383.

70. Sarkany RP. 2001. The management of porphyria cutanea tarda. *Clin Exp Dermatol* 26: 225–232.

71. Elder GH. 1999. Alcohol intake and porphyria cutanea tarda. *Clin Dermatol* 17: 431–436.

72. Moran-Jimenez MJ, Ged C, Romana M, Enriquez De Salamanca R, Taïeb A, Topi G, D'Alessandro L, de Verneuil H. 1996. Uroporphyrinogen decarboxylase: Complete human gene sequence and molecular study of three families with hepatoerythropoietic porphyria. *Am J Hum Genet* 58: 712–721.

73. Kordac V. 1972. Frequency of occurrence of hepatocellular carcinoma in patients with porphyria cutanea tarda in long-term follow-up. *Neoplasma* 19: 135–139.

74. Cortes JM, Oliva H, Paradinas FJ, Hernandez-Guio C. 1980. The pathology of the liver in porphyria cutanea tarda. *Histopathology* 4: 471–485.

75. Pischik E, Kauppinen R. 2009. Neurological manifestations of acute intermittent porphyria. *Cell Mol Biol (Noisy-le-grand)* 55: 72–83.

76. Herrick AL, McColl KE. 2005. Acute intermittent porphyria. *Best Pract Res Clin Gastroenterol* 19: 235–249.

77. Kauppinen R, Mustajoki P. 1992. Prognosis of acute porphyria: Occurrence of acute attacks, precipitating factors, and associated diseases. *Medicine (Baltimore)* 71: 1–13.

78. Bylesjo I, Wikberg A, Andersson C. 2009. Clinical aspects of acute intermittent porphyria in northern Sweden: A population-based study. *Scand J Clin Lab Invest* 69: 612–618.

79. Kochar DK, Pal M, Kochar SK, Vyas A, Kochar A, Bindal D, Agrawal RP. 2007. Acute intermittent porphyria presenting with neurological emergency: Review of six cases. *Neurol India* 55: 413–415.

80. Palmer KM. 2006. Abdominal pain due to acute intermittent porphyria: When is the sound of hoof-beats not horses, but zebras? A case report. *Dimens Crit Care Nurs* 25: 103–109.

81. Waldenstrom J. 1957. The porphyrias as inborn errors of metabolism. *Am J Med* 22: 758–773.

82. Goldberg A. 1959. Acute intermittent porphyria: A study of 50 cases. *Q J Med.* 28: 183–209.

83. Stein JA, Tschudy DP. 1970. Acute intermittent porphyria. A clinical and biochemical study of 46 patients. *Medicine* 49: 1–16.

84. Asselbergs FW, Kremer Hovinga TK, Bouwsma C, van Ingen J. 2009. Acute intermittent porphyria as a cause of respiratory failure: Case report. *Am J Crit Care* 18: 180, 78–79.

85. Granick S, Van Den Schreieck HG. 1955. Porphobilinogen and delta-aminolevulinic acid in acute porphyria. *Pros Soc Exp Biol Med* 88: Starts at 270.

86. Robert TL, Varella L, Meguid MM. 1994. Nutrition management of acute intermittent porphyria. *Nutrition* 10: 551–555; quiz 6–7.

87. Atamna H. 2004. Heme, iron, and the mitochondrial decay of ageing. *Ageing Res Rev* 3: 303–318.

88. Zhu Y, Hon T, Ye W, Zhang L. 2002. Heme deficiency interferes with the Ras-mitogen-activated protein kinase signaling pathway and expression of a subset of neuronal genes. *Cell Growth Differ* 13: 431–439.

89. Sengupta A, Hon T, Zhang L. 2005. Heme deficiency suppresses the expression of key neuronal genes and causes neuronal cell death. *Brain Res Mol Brain Res* 137: 23–30.

90. Dwyer BE, Smith MA, Richardson SL, Perry G, Zhu X. 2009. Down-regulation of aminolevulinate synthase, the rate-limiting enzyme for heme biosynthesis in Alzheimer's disease. *Neurosci Lett* 460: 180–184.

91. Tenhunen R, Marver HS, Schmid R. 1969. Microsomal heme oxygenase. Characterization of the enzyme. *J Biol Chem* 244: 6388–6394.

92. Chowdhury JR, Wolkoff AW, Chowdhury NR, Arias IM. 2009. Hereditary Jaundice and Disorders of Bilirubin Metabolism. In *The Metabolic and Molecular Bases of Inherited Disease*, eds. CR Scriver, AL Beaudt, WS Sly, D Valle, C Barton, KW Kinzler, B Vogelstein, Chapter 125, pp. 1–90. New York: The McGraw-Hill Companies, Inc.

93. Li C, Stocker R. 2009. Heme oxygenase and iron: From bacteria to humans. *Redox Rep* 14: 95–101.

94. Jackson AH, Kenner W. 1968. Recent developments in porphyrin chemistry. In *Porphyrins and related compounds*, pp. 5: London: Academic Press.

95. Bissell DM, Hammaker L, Schmid R. 1972. Liver sinusoidal cells. Identification of a Subpopulation for Erythrocyte Catabolism. *J Cell Biol* 54: 107–119.

96. Maines MD, Trakshel GM, Kutty RK. 1986. Characterization of two constitutive forms of rat liver microsomal heme oxygenase. Only one molecular species of the enzyme is inducible. *J Biol Chem* 261: 411–419.

97. McCoubrey WK, Jr., Huang TJ, Maines MD. 1997. Heme oxygenase-2 is a hemoprotein and binds heme through heme regulatory motifs that are not involved in heme catalysis. *J Biol Chem* 272: 12568–12574.

98. Wagener FA, Volk HD, Willis D, Abraham NG, Soares MP, Adema GJ, Figdor CG. 2003. Different faces of the heme-heme oxygenase system in inflammation. *Pharmacol Rev* 55: 551–571.

99. Sikorski EM, Hock T, Hill-Kapturczak N, Agarwal A. 2004. The story so far: Molecular regulation of the heme oxygenase-1 gene in renal injury. *Am J Physiol Renal Physiol* 286: F425–F441.

100. Schipper HM. 2004. Heme oxygenase expression in human central nervous system disorders. *Free Radic Biol Med* 37: 1995–2011.

101. Schipper HM, Chertkow H, Mehindate K, Frankel D, Melmed C, Bergman H. 2000. Evaluation of heme oxygenase-1 as a systemic biological marker of sporadic AD. *Neurology* 54: 1297–1304.

102. Schipper HM, Cisse S, Stopa EG. 1995. Expression of heme oxygenase-1 in the senescent and Alzheimer-diseased brain. *Ann Neurol* 37: 758–768.

103. Sassa S, Kappas A, Bernstein SE, Alvares AP. 1979. Heme biosynthesis and drug metabolism in mice with hereditary hemolytic anemia. Heme oxygenase induction as an adaptive response for maintaining cytochrome P-450 in chronic hemolysis. *J Biol Chem* 254: 729–735.

104. Immenschuh S, Ramadori G. 2000. Gene regulation of heme oxygenase-1 as a therapeutic target. *Biochem Pharmacol* 60: 1121–1128.

105. Lee PJ, Jiang BH, Chin BY, Iyer NV, Alam J, Semenza GL, Choi AM. 1997. Hypoxia-inducible factor-1 mediates transcriptional activation of the heme oxygenase-1 gene in response to hypoxia. *J Biol Chem* 272: 5375–5381.

106. Takahashi S, Takahashi Y, Yoshimi T, Miura T. 1998. Oxygen tension regulates heme oxygenase-1 gene expression in mammalian cell lines. *Cell Biochem Funct* 16: 183–193.

107. Motterlini R, Foresti R, Bassi R, Calabrese V, Clark JE, Green CJ. 2000. Endothelial heme oxygenase-1 induction by hypoxia. Modulation by inducible nitric-oxide synthase and S-nitrosothiols. *J Biol Chem* 275: 13613–13620.

108. Yang ZZ, Zou AP. 2001. Transcriptional regulation of heme oxygenases by HIF-1alpha in renal medullary interstitial cells. *Am J Physiol Renal Physiol* 281: F900–F908.

109. Shibahara S, Nakayama M, Kitamuro T, Udono-Fujimori R, Takahashi K. 2003. Repression of heme oxygenase-1 expression as a defense strategy in humans. *Exp Biol Med (Maywood)* 228: 472–473.

110. Kacimi R, Chentoufi J, Honbo N, Long CS, Karliner JS. 2000. Hypoxia differentially regulates stress proteins in cultured cardiomyocytes: Role of the p38 stress-activated kinase signaling cascade, and relation to cytoprotection. *Cardiovasc Res* 46: 139–150.

111. Kitamuro T, Takahashi K, Ogawa K, Udono-Fujimori R, Takeda K, Furuyama K, Nakayama M, Sun J, Fujita H, Hida W, Hattori T, Shirato K, Igarashi K, Shibahara S. 2003. Bach1 functions as a hypoxia-inducible repressor for the heme oxygenase-1 gene in human cells. *J Biol Chem* 278: 9125–9133.

112. Vitek L, Ostrow JD. 2009. Bilirubin chemistry and metabolism; harmful and protective aspects. *Curr Pharm Des* 15: 2869–2883.

113. Berk PD, Howe RB, Bloomer JR, Berlin NI. 1969. Studies of bilirubin kinetics in normal adults. *J Clin Invest* 48: 2176–2190.

114. London IM, West R, Shemin D, D. R. 1949. On the origin of bile pigment in normal man. *J. Biol Chem* 184: 351–358.

115. Gordon ER, Goresky CA, Chang TH, Perlin AS. 1976. The isolation and characterization of bilirubin diglucuronide, the major bilirubin conjugate in dog and human bile. *Biochem J* 155: 477–486.

116. Bosma PJ, Seppen J, Goldhoorn B, Bakker C, Oude Elferink RP, Chowdhury JR, Chowdhury NR, Jansen PL. 1994. Bilirubin UDP-glucuronosyltransferase 1 is the only relevant bilirubin glucuronidating isoform in man. *J Biol Chem* 269: 17960–17964.

117. Ollinger R, Wang H, Yamashita K, Wegiel B, Thomas M, Margreiter R, Bach FH. 2007. Therapeutic applications of bilirubin and biliverdin in transplantation. *Antioxid Redox Signal* 9: 2175–2185.

118. Ollinger R, Yamashita K, Bilban M, Erat A, Kogler P, Thomas M, Csizmadia E, Usheva A, Margreiter R, Bach FH. 2007. Bilirubin and biliverdin treatment of atherosclerotic diseases. *Cell Cycle* 6: 39–43.

119. Stocker R, Yamamoto Y, McDonagh AF, Glazer AN, Ames BN. 1987. Bilirubin is an antioxidant of possible physiological importance. *Science* 235: 1043–1046.

120. Halliwell B, Gutteridge JM. 1990. The antioxidants of human extracellular fluids. *Arch Biochem Biophys* 280: 1–8.

121. Dennery PA, McDonagh AF, Spitz DR, Rodgers PA, Stevenson DK. 1995. Hyperbilirubinemia results in reduced oxidative injury in neonatal Gunn rats exposed to hyperoxia. *Free Radic Biol Med* 19: 395–404.

122. Kirk JM. 2008. Neonatal jaundice: A critical review of the role and practice of bilirubin analysis. *Ann Clin Biochem* 45: 452–462.

123. Maisels MJ, McDonagh AF. 2008. Phototherapy for neonatal jaundice. *N Engl J Med* 358: 920–928.

124. Kaplan M, Hammerman C. 2005. Understanding severe hyperbilirubinemia and preventing kernicterus: adjuncts in the interpretation of neonatal serum bilirubin. *Clin Chim Acta* 356: 9–21.

125. Gourley GR. 1997. Bilirubin metabolism and kernicterus. *Adv Pediatr* 44: 173–229.

126. Scharschmidt BF, Gollan JL. 1979. Current concepts of bilirubin metabolism and hereditary hyperbilirubinemia. *Prog Liver Dis* 6: 187–212.

127. Wolkoff AW. 1983. Inheritable disorders manifested by conjugated hyperbilirubinemia. *Semin Liver Dis* 3: 65–72.

128. Huang CS. 2005. Molecular genetics of unconjugated hyperbilirubinemia in Taiwanese. *J Biomed Sci* 12: 445–450.

129. Jansen PL. 1999. Diagnosis and management of Crigler-Najjar syndrome. *Eur J Pediatr* 158 Suppl 2: S89–S94.

130. Sampietro M, Iolascon A. 1999. Molecular pathology of Crigler-Najjar type I and II and Gilbert's syndromes. *Haematologica* 84: 150–157.

131. Ritter JK, Yeatman MT, Ferreira P, Owens IS. 1992. Identification of a genetic alteration in the code for bilirubin UDP-glucuronosyltransferase in the UGT1 gene complex of a Crigler-Najjar type I patient. *J Clin Invest* 90: 150–155.

132. Bosma PJ, Chowdhury NR, Goldhoorn BG, Hofker MH, Oude Elferink RP, Jansen PL, Chowdhury JR. 1992. Sequence of exons and the flanking regions of human bilirubin-UDP-glucuronosyltransferase gene complex and identification of a genetic mutation in a patient with Crigler-Najjar syndrome, type I. *Hepatology* 15: 941–947.

133. Zimniak P. 1993. Dubin-Johnson and Rotor syndromes: Molecular basis and pathogenesis. *Semin Liver Dis* 13: 248–260.

134. Luketic VA, Shiffman ML. 2004. Benign recurrent intrahepatic cholestasis. *Clin Liver Dis* 8: 133–149, vii.

Chapter 3

HEME

An Ingenious Regulator of Gene Transcription

Li Zhang

Overview

One important regulatory role of heme is to control the expression of diverse genes by modulating the activity of transcriptional regulators. In mammals, heme regulates the transcription of many genes that are crucial for many fundamental processes, including circadian rhythmicity, pancreatic development, and the generation of red blood cells. In yeast, heme regulates the transcription of many genes encoding functions required for respiration and for controlling oxidative damage. Two well-studied examples of heme-regulated transcription factors are the yeast heme activator protein Hap1 and the mammalian transcriptional repressor Bach1. In yeast, Hap1 activates the transcription of many genes in response to increased intracellular heme concentrations or oxygen levels. In mammalian cells, at low cellular heme concentrations, Bach1 represses the transcription of many genes, including those encoding globin chains. When heme concentration increases, heme inactivates Bach1 and allows the repressed genes to be derepressed and activated. In this chapter, I explain in detail how heme controls the activity of these two important regulators Hap1 and Bach1.

Heme can control the activity of diverse transcriptional regulators

Organisms and tissues exhibit different capabilities and characteristics largely due to the different proteins that are expressed in their cells.

Gene transcription is the first step in the process of gene expression — the process by which the synthesis of a functional gene product, usually a protein, is accomplished based on the information from a gene (1). Gene transcription, or RNA synthesis, allows the synthesis of an equivalent RNA (ribonucleic acid) copy of a sequence of DNA (deoxyribonucleic acid). The DNA sequence is read, and RNA is synthesized by an enzyme called RNA polymerase. The transcription of various genes is controlled via adjusting the activity of RNA polymerase. There are many ways to do that. One of the most common ways is by the action of regulatory proteins that can modulate the activity of RNA polymerase (1). There are two classes of regulatory proteins: positive regulators that enhance the activity of RNA polymerase, and negative regulators or repressors that diminish the activity of RNA polymerase. Such regulatory proteins are often responsive to intracellular and extracellular signals or cues.

Heme is a regulator that can control gene transcription by binding to and modulating the activity of regulatory proteins in diverse organisms (2–5). The activities of such regulatory proteins often rise and fall with intracellular heme concentration. Such proteins include the yeast transcriptional regulator Hap1 (2, 6), the mammalian transcriptional repressor Bach1 (7–10), and the mammalian nuclear receptor Rev-erbα (11–13). How heme controls the activity of Hap1 and Bach1 to regulate gene transcription is the best investigated and best understood. I will therefore explain these two cases in detail in this chapter. These cases should provide a clear idea of how heme can work ingeniously to control gene expression in yeast and mammals.

Heme serves as a secondary messenger of oxygen in yeast

Heme biosynthesis in yeast is achieved by the actions of the same eight enzymes as those in mammals (14–16) described in Chapter 2. Two of these enzymes, coproporphyrinogen III oxidase and protoporphyrinogen oxidase, use oxygen as a substrate (14). In addition, Hon *et al.* (17) found that the activity of the enzyme that catalyzes the last step of heme synthesis, ferrochelatase, is also under the control of oxygen. All these effects of oxygen on heme synthesis allow intracellular heme concentration to be coordinated with the oxygen concentration in the environment, at the oxygen range of 0.2–2 micromolar (17). At oxygen concentrations higher than 2 micromolar, heme is synthesized at high levels constitutively. This correlation between heme concentration and oxygen level affords heme a central role in mediating the effect of oxygen on gene expression. As such, the activities of proteins and enzymes can be controlled by heme, instead of oxygen.

Heme can control the expression of a high number of genes in yeast

Although heme can serve as a secondary messenger of oxygen in yeast, its role is not limited in mediating oxygen regulation. Data from previous microarray gene expression profiling analyses showed that heme controls, directly or indirectly, the expression of a large number of genes in the yeast *Sacchromyces cerevisiae*, a budding yeast used in brewing and baking (Fig. 1). A fraction of the heme-regulated (induced and suppressed) genes are also regulated by oxygen, showing that heme indeed mediates oxygen regulation of a group of genes (see Figs. 1a and 1b). Fig. 1a shows that $324\,(303+21)$ genes are induced or activated by both heme and oxygen, while $571\,(568+3)$ and $637\,(620+17)$ genes are induced only by oxygen and heme, respectively. Fig. 1b shows that $139\,(112+27)$ genes are suppressed by both heme and oxygen, while $344\,(334+10)$ and $341\,(337+4)$ genes are suppressed only by oxygen and heme, respectively. Hap1 is one of the regulators mediating both oxygen and heme regulation. As shown in Fig. 1, it mediates the activation of 27 genes activated by both heme and oxygen, and the suppression of 21 genes suppressed by both heme and oxygen. In addition, Hap1 can activate eight (Fig. 1a) and can repress nine other genes (Fig. 1b).

Notably, a substantial number of heme-regulated genes are not regulated by oxygen, showing that heme can also regulate the expression of genes that are not affected by oxygen. Computational analyses further identified potential regulators that may mediate oxygen and/or heme regulation of gene expression (Fig. 2). The results shown in Fig. 2 suggest that a group of regulators $(13+5)$ can mediate both oxygen and heme regulation of gene expression. A group of regulators, $18\,(16+2)$, appear to mediate only oxygen regulation, whereas another group of regulators, $7\,(5+2)$, appear to mediate only heme regulation. The regulation of gene expression by Hap1 appears to involve only regulators that mediate heme and/or oxygen regulation. Fig. 2 shows that there is no regulator that

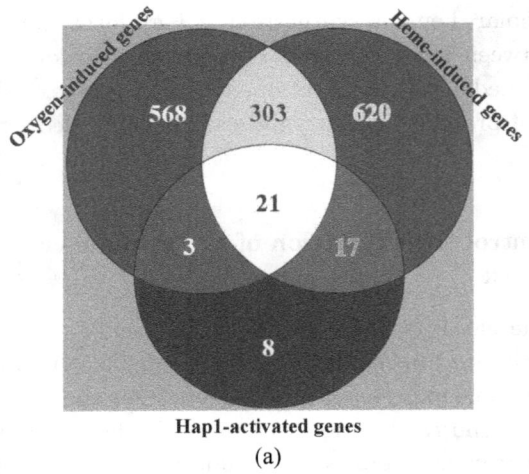

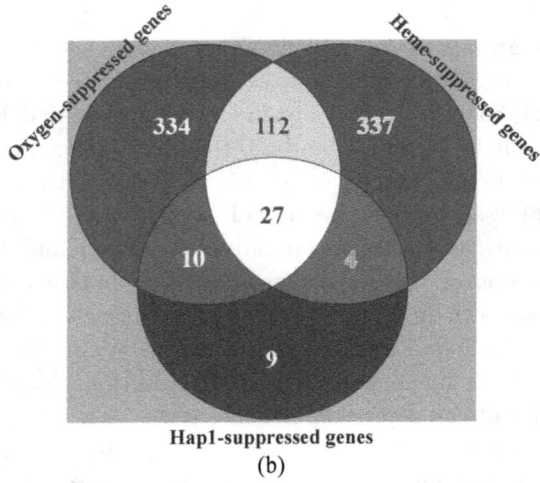

Fig. 1. Heme regulates the expression of a high number of genes in yeast. (a) A Venn diagram illustrating the numbers of genes induced by oxygen, heme and Hap1. (b) A Venn diagram illustrating the numbers of genes suppressed by oxygen, heme and Hap1.

may mediate only the regulation of gene expression by Hap1, but not by heme or oxygen. These results show that both oxygen and heme control the expression of a high number of genes, while Hap1 is a key regulator mediating oxygen and heme regulation.

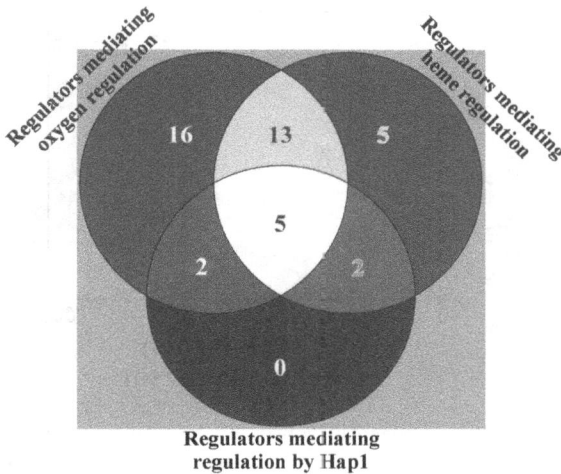

Fig. 2. A Venn diagram showing the numbers of regulators mediating oxygen regulation, heme regulation, and Hap1 regulation of gene expression in yeast.

The heme activator protein Hap1 is a key regulator of gene transcription in response to fluctuations in oxygen and heme levels

Hap1 is a large yeast protein containing 1483 amino acid residues (18). Hap1 activity is controlled stringently by heme concentration (Fig. 3). When oxygen or heme level is high, it activates the transcription of genes that encode functions required for respiration and for controlling oxidative damage (6, 18–23). These include cytochromes, catalase, and yeast hemoglobin (6, 23). Interestingly, these proteins and enzymes often use heme as an essential cofactor or prosthetic group (24). The whole process forms a continuous circle with heme and Hap1 in control (Fig. 4). When cells sense the presence of a substantial amount of oxygen, they make heme in the mitochondria, the organelle that enables respiration. Then, heme goes to the nucleus, the organelle where DNA is stored and RNA is synthesized, to activate Hap1. Subsequently, genes that encode functions required for respiration and for controlling oxidative damage are transcribed, and proteins are made. These proteins then use heme to form functional proteins and enzyme complexes, which in turn transport and use oxygen. These proteins and enzymes may function in various organelles, including mitochondria, the cytoplasm and others (Fig. 4).

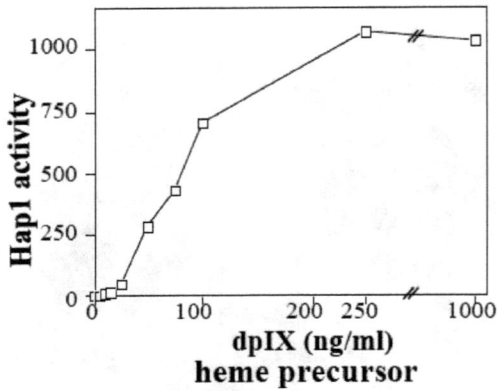

Fig. 3. Hap1 activity is stringently controlled by heme concentration. The plot shows the correlation between Hap1 transcriptional activity with the levels of heme. Heme levels are controlled by addition of the heme precursor, deuteroporphyrin IX (dpIX).

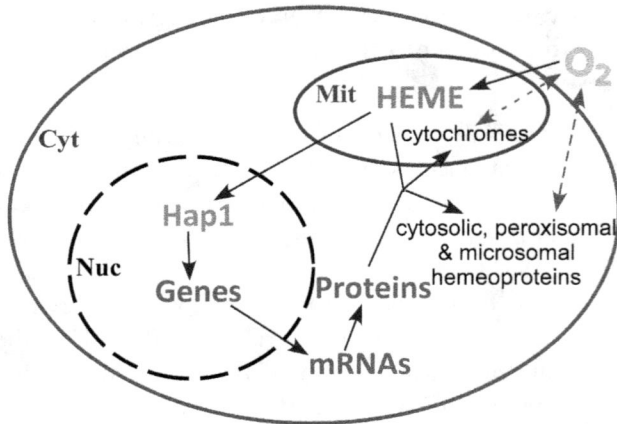

Fig. 4. The heme-Hap1 regulatory loop in the budding yeast. When cells are grown aerobically with high levels of oxygen, heme is synthesized in the mitochondria. Heme is transported to the nucleus, where it activates Hap1. Activated Hap1 in turn stimulates the transcription of the genes that encode functions required for respiration and for controlling oxidative damage. Transcripts or mRNAs are then translated to make proteins. Many of these proteins, like cytochromes and other hemoproteins, then incorporate heme and form functional protein and enzymes complexes. These complexes in turn often act to transport or use oxygen.

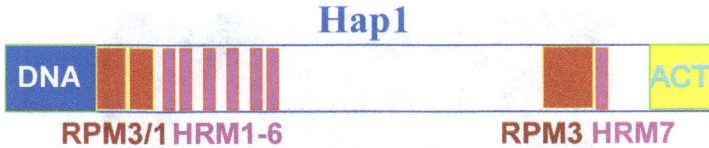

Fig. 5. The domain structure of Hap1 protein. Shown here are the domains or regions of the Hap1 protein that mediate DNA binding (DNA), heme regulation (RPMs and HRMs), and transcriptional activation (ACT). RPM1–3 mediates the repression of Hap1 activity when cellular heme concentration is low. HRMs can bind heme. In Hap1, only one HRM, HRM7, is necessary for conferring the activation of Hap1 activity by heme.

Hap1 possesses the typical characteristics of transcriptional activators. It has a region called the DNA-binding domain that allows Hap1 to bind to DNA at the N-terminus (18) (Fig. 5). The Hap1 DNA-binding domain contains a C6 zinc cluster motif that is found in the yeast Gal4 family of transcriptional regulators (25). The DNA-binding domains of this family of regulators also contain a dimerization element, which allows them to dimerize and bind to DNA as dimers (26–29). The C6 zinc cluster motif recognizes and binds to a CGG triplet in DNA (27, 28, 30) (see Fig. 6). Dimers of these transcriptional regulators bind to DNA sites containing two CGG triplets (27, 28, 30). Hap1 is unique among the Gal4 family of C6 zinc cluster proteins, because the Hap1 dimer binds to an asymmetric DNA site containing a direct repeat of CGG triplets, instead of symmetrical DNA sites containing an inverted repeat of two CGG triplets (31–33). The structure of the Hap1-DNA complex (33) is shown in Fig. 6.

At the C-terminus, Hap1 also has a region called the activation domain that allows Hap1 to make contact, directly or indirectly, with RNA polymerase to activate transcription (Fig. 5) (18). Additionally, in the interweaving region, a series of amino acid motifs and modules exist to mediate heme regulation of Hap1 activity (34, 35). These include the HRMs (heme regulatory or responsive motifs). There are seven of them in Hap1 (Fig. 5): HRM1–6 near the DNA binding domain and HMR7 near the activation domain. They are required for heme binding to Hap1 and for promoting Hap1 to bind to DNA and activate transcription. Notably, only HRM7 is necessary and sufficient for mediating heme binding and activation of Hap1. Besides HRMs, there are three important modules called RPMs (repression modules) (Fig. 5): two near the DNA binding

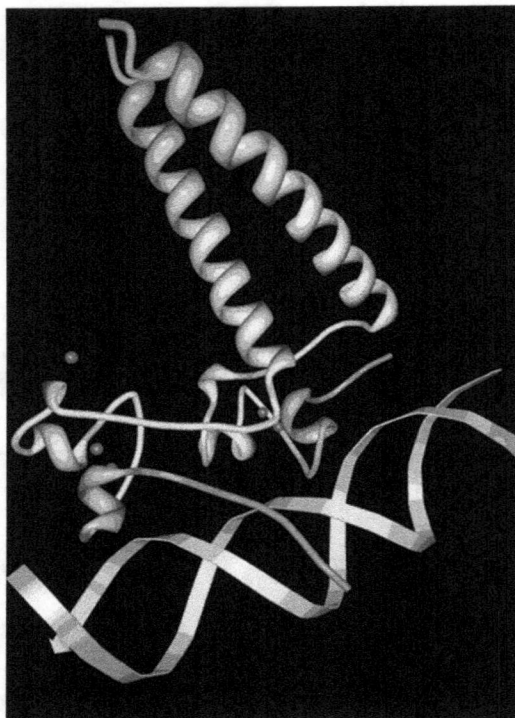

Fig. 6. The structure of the Hap1-DNA complex. The Hap1 fragment, containing residues 55–135 in the Hap DNA-binding domain, is used in the structural analysis. The image is generated based on the data from the Protein Data Bank.

domain and one near the activation domain (34, 35). They are responsible for keeping Hap1 inactive when the heme concentration is low. The existence of two distinct classes of Hap1 elements mediating heme regulation is also a distinctive characteristic of Hap1. This enables heme to regulate Hap1 stringently via a two-tier regulatory mechanism (6).

Although Hap1 activity fluctuates with oxygen levels, this is mediated completely via heme. Oxygen is not needed to activate Hap1 if heme is present (17). Likewise, Hap1 is inactivated when the heme concentration is low, even if the oxygen level in the environment is high. This kind of regulators, including Hap1 (6) and Bach1 (9), are different from other kinds of regulators, such as the neuronal transcriptional regulator, NPAS2 (36), which use heme as a prosthetic group to bind gases such as oxygen and carbon monoxide. Importantly, the regulation of Hap1 activity by heme also requires the action of other kinds of proteins known as molecular chaperones.

Heme regulation of Hap1 transcription-activating activity requires the proper functioning of Hsp90 and Hsp70 molecular chaperones

Although Hap1 contains motifs that can bind to heme, and these motifs are critical for the activation of Hap1 by heme, Hap1 by itself cannot respond to heme properly. Work in the Zhang lab identified two classes of proteins, which are known as molecular chaperones (37–40). As indicated by their names, they play essential roles in helping other proteins function. Particularly, Hsp70 and Hsp90 (with approximate molecular masses of 70 and 90 kDa, respectively) molecular chaperones are known to be critical for proper functioning of many regulatory or signaling proteins, such as steroid receptors, in diverse organisms (41–57). In the same vein, they are also required for the proper functioning of Hap1, a *de facto* heme receptor. In cells with limiting levels of Hsp90, Hap1 does not become fully activated even at high heme concentrations (39). In cells with limiting levels of Hsp70, Hap1 becomes fully active even at low heme concentrations (37–40).

It appears that Hsp70 chaperones act together with RPMs to ensure that Hap1 is inactive in cells with low concentrations of heme, whereas Hsp90 chaperones act with HRMs to allow Hap1 to bind heme and to be activated (37–40, 58). Such a mechanism affords two tiers of control of Hap1 activity, so that a precise control of Hap1 activity by heme is achieved. Based on the work on molecular chaperones and Hap1 motifs and modules, a two-tier regulatory mechanism was suggested to explain how heme regulates Hap1 activity (Fig. 7) (40).

Very likely, newly synthesized Hap1 is bound immediately by Hsp70 molecular chaperons and their cochaperones Ydj1 and Sro9. As a result, Hap1 can be folded properly and be repressed. In this repressed state, Hap1 also adopts an activatable conformation, so that it can be readily activated when the heme concentration increases (Fig. 7). Ydj1 assists Hsp70 in a variety of process (59, 60), whereas Sro9 appears to act specifically in heme regulation of Hap1 (38, 61). RPMs of Hap1 likely make contacts with Hsp70. Hsp90 may also associate with this Hap1-Hsp70 complex transiently and dynamically. When the heme concentration increases, Hsp90 binds to the Hap1-Hsp70 complex much more strongly,

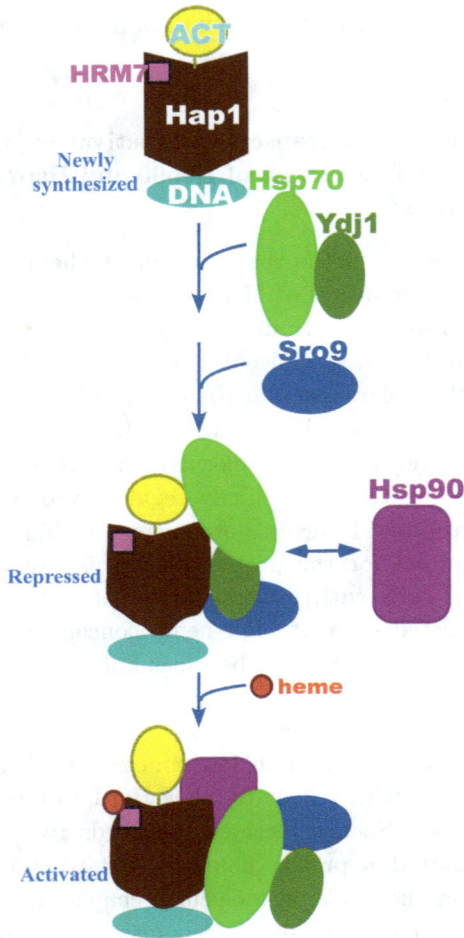

Fig. 7. A cartoon illustrating the steps and complexes that may be involved in mediating the regulation of Hap1 activity by heme. Newly synthesized Hap1 is postulated to be bound, likely cotranslationally, by Hsp70 and cochaperone proteins Ydj1 and Sro9, forming a repressed complex. Hsp90 may transiently and dynamically interact with this repressed complex. When the intracellular heme concentration increases, heme binds to HRM7 and changes the conformation of Hap1 and the whole complex. Consequently, Hsp90 binds to the complex strongly, causing Hap1 to be fully activated and turning on the transcription of various genes. For simplicity, all proteins are shown as monomers, although Hap1 and Hsp90 are known to function as dimers.

and likely causes a conformational change in the complex (40, 58). As a result, heme binds to Hap1 via HRM7, leading to the activation of Hap1 transcription-activating activity (Fig. 7) (39, 58).

If the level of Hsp90 is low or if Hsp90 is mutated, this conformational change can be inhibited. Consequently, Hap1 remains inactive even at high levels of heme (39, 40, 58). Interestingly, the mere association of Hsp90 with Hap1 does not allow Hap1 to be activated by heme. For example, when the Gly amino acid residue at position 313 in Hsp90 is mutated to Asp, the mutant can still interact with Hap1, but is not able to support heme activation of Hap1 (39). Likewise, when the HRM7 motif in Hap1 is mutated, the Hap1 mutant can still interact with Hsp90, but cannot be activated by heme (58). These data show that conformational changes are necessary for Hsp90 function and for heme activation of Hap1 (39, 40, 58). The Hap1-heme regulatory system is arguably the best understood example of how heme controls the activity of regulatory or signaling proteins. It provides a paradigm for studying other heme-regulated proteins and how molecular chaperones may aid the regulation of regulatory or signaling proteins (39).

Bach1: A heme-dependent transcriptional regulator in mammalian cells

If the Hap1-heme regulatory system is an example of how heme can activate the activity of positive transcriptional regulators, Bach1 provides an example of how heme can control the activity of negative transcriptional regulators, or repressors (7–10, 62–64). Bach1 is the first mammalian transcriptional factor found to bind heme and to be regulated by heme concentration (9, 10). Bach1 is expressed in diverse mammalian cells, ranging from erythroid cells to skin and lung cells (7, 65–68). The target genes of Bach1 include α- and β-globin genes and the heme oxygenase gene (2, 3, 7). The globin genes encode the globin chains, which bind to heme and form oxygen-transport hemoglobin in mammals (69). The heme oxygenase gene encodes the enzyme that degrades heme, as described in Chapter 2. It converts heme to free iron, carbon monoxide (a neuromodulator), and biliverdin (70–72). In mammals, biliverdin is further transformed into bilirubin, an antioxidant and anti-inflammatory agent (73).

The transcription of globin genes and the heme oxygenase gene is induced by heme (2, 3). These genes contain a DNA site called MARE in their enhancers — the DNA elements that mediate the enhancement of transcription from a distance (7, 8). Bach1 forms heterodimers with a class of proteins called small Maf proteins. The Bach1-Maf heterodimer binds to MARE and causes the repression of the genes (9, 74). When the intracellular heme concentration increases, the Bach1-Maf heterodimer is displaced by activators, which then stimulate the transcription of the genes containing MARE (9, 74). Thus, the regulation of Bach1 activity by heme is a crucial mechanism for regulating the expression of genes containing MARE in mammalian cells.

Bach1, like Hap1, binds to heme via HRMs

Bach1 in human cells contains 739 amino acid residues (8). It also contains multiple domains (Fig. 8): the BTB domain that mediates protein-protein interaction, the CLS domain that facilitates cytoplasmic accumulation, and the basic Zip domain that facilitates nuclear accumulation and DNA binding (7–9, 74–76). Additionally, Bach1, like Hap1, contains HRMs (Fig. 8) (2, 77). Six HRMs are found in Bach1, although only HRM3–6 appears to be important for heme regulation of Bach1 (2, 77). Recent studies suggest that five molecules of heme bind to Bach1 (77). The chemistry of heme binding to HRMs is quite simple (78). As shown in Fig. 9, HRMs in Hap1 and Bach1, like HRMs in other proteins, invariably contain the CP residues (2, 77) (Fig. 9). Each HRM containing one CP motif can bind one molecule of heme. The Cys residue coordinates the Fe ion in heme and is crucial for heme binding. Mutation of the Cys residues in either Hap1 or Bach1 abolishes heme binding *in vitro*, and heme regulation *in vivo*.

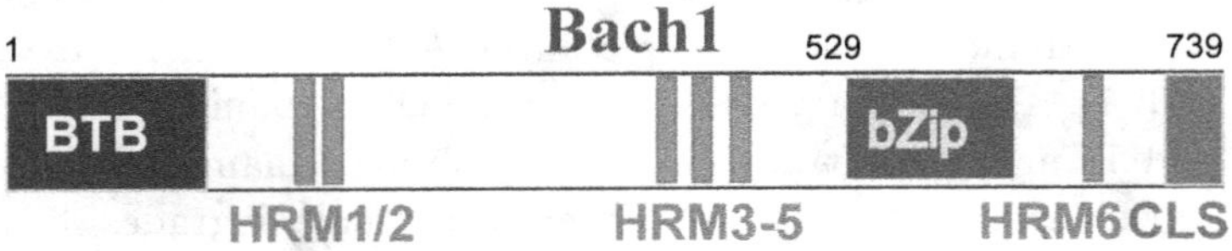

Fig. 8. The domain structure of Bach1 protein. Shown here is the Bach1 BTB domain that mediates protein-protein interaction, the HRMs that bind to heme and mediate heme regulation, the CLS domain that facilitates cytoplasmic accumulation, and the basic Zip domain that facilitates nuclear accumulation and DNA binding. Only HRM3–6 are required for the regulation of Bach1 activity by heme.

Hap1

HRM1	KC**P**INH
2	KC**P**VDH
3	RC**P**VDH
4	RC**P**VDH
5	KC**P**VDH
6	RC**P**IDH
7	KC**P**VYQ

Bach1

HRM1	LC**P**KYR
2	QC**P**TEK
3	EC**P**WLG
4	NC**P**FIS
5	PC**P**YAC
6	DC**P**LSF

Fig. 9. Amino acid sequences of heme-responsive motifs (HRMs) in Hap1 and Bach1. The CP residues are invariably present in all motifs.

Heme affects Bach1 at three levels

Initially, it was found that heme binds to Bach1 and regulates its DNA-binding activity (9). Specifically, heme binds to Bach1, via the C-terminal region containing HRM3–6 (Fig. 8), and inhibits the DNA-binding activity of the Bach1-Maf heterodimer. When heme binds to Bach1, the Bach1-Maf heterodimer dissociates from DNA, and causes the derepression and induction of a host of genes with MAREs, such as the heme oxygenase gene. Subsequently, it was shown that heme also triggers the export of Bach1 out of the nucleus (75). Heme-induced Bach1 nuclear export requires HRM3 and HMR4, but not HRM5 and HRM6. A protein called Crm1 or exportin 1 assists the nuclear export of Bach1 (75, 76).

Still, more recent work suggests that heme regulates the degradation of Bach1 (79). Micromolar concentrations of heme were shown to cause a significant reduction of the Bach1 protein level in diverse cells, including erythroid and fibroblast cells. The degradation of Bach1 is performed by sophisticated protein degradation machinery called proteasomes (80–82). To be degraded, a protein needs to be first ubiquitinated (83, 84). This ubiquitination involves the covalent attachment of one or more ubiquitin molecules to the protein. Ubiquitin is a small protein containing 76 amino acid residues (85). Once ubiquitinated, proteins may be degraded by the proteasome degradation machinery (80–82). In the case of Bach1, a protein called HOIL-1 facilitates its ubiquitination and degradation (79, 86).

It appears that HRM3–6 is important for heme-induced, HOIL-1-mediated degradation of Bach1 (79). The Bach1 region containing HRM3–5 may contact HOIL-1 directly. Interestingly, HOIL-1 also facilitates

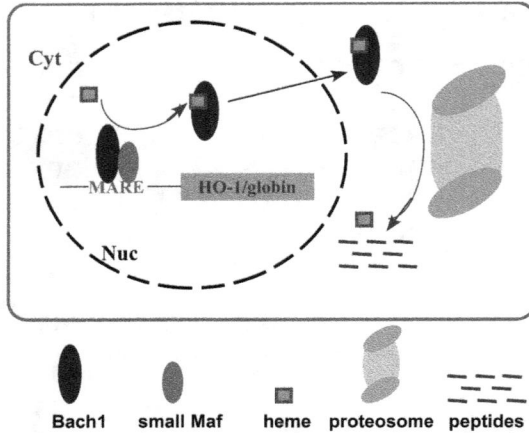

Fig. 10. A cartoon illustrating the regulation of Bach1 activity by heme. When the intracellular heme concentration is low, genes containing the MARE sequence are bound by the Bach1-Maf dimer in the nucleus (Nuc), and are repressed. When the intracellular heme concentration increases, heme binds to Bach1, causing it to dissociate from MARE, and relocate to the cytoplasm (Cyt). Relocated Bach1 in the cytoplasm is then degraded into peptides by the proteasome machinery. The dissociation of Bach1 from MARE, its relocalization, and degradation all require heme binding.

heme-induced degradation of another important regulatory protein, called IRP2 (iron regulatory protein 2). It is a regulator of iron homeostasis in mammals. Heme binds to it through an HRM in IRP2 and promotes degradation (87).

Taken together, the data in the current literature show that heme regulates Bach1 activity via three different modes (Fig. 10) (9, 75, 77, 79). When the intracellular heme concentration is low, Bach1-Maf heterodimer binds to genes containing MARE, such as the globin genes and the heme oxygenase gene. These genes are repressed by Bach1. When the intracellular heme concentration increases, heme binds to Bach1 via HRMs 3–6, causing the Bach1-Maf heterodimer to dissociate from MARE. Then, Crm1 (exportin 1) becomes associated with Bach1. This association facilitates the export of Bach1 out of the nucleus. Subsequently, HOIL-1 recognizes and binds to Bach1, causing it to be ubiquitinated and degraded by the proteasome. All three steps, dissociation from DNA, nuclear export, and degradation, require the binding of heme to HRMs.

Perspectives

Besides Bach1 and Hap1, heme has the capacity to regulate the activity of other transcription factors. A recent computational search shows that over 50 transcription activators contain the potential heme-binding CP motifs (Arvey *et al.*, unpublished data). Experimental evidence also suggests that other transcription factors can be regulated by heme. For example, the transcriptional regulator Ptf1a is regulated by heme in zebra fish (88). Notably, the mammalian nuclear receptor Rev-erbα, a transcriptional regulator that coordinates metabolic and circadian pathways, binds to heme and is regulated by heme (11–13). Intriguingly, heme binds to Rev-erbα via a histidine residue, although it contains an HRM-like motif with the CP residues. What, if any role, this HRM may play in heme binding and regulation, is not clear. Although the His residue is critical for heme binding and regulation, it has not been shown that it is sufficient. Thus, there may still be a potential role for HRM to play in heme regulation of Rev-erbα. There is little doubt that more transcription factors will be shown to be regulated by heme, as more studies are performed.

References

1. Lewin B. 2007. *Genes IX*. Jones and Bartlett Learning.
2. Mense SM, Zhang L. 2006. Heme: A versatile signaling molecule controlling the activities of diverse regulators ranging from transcription factors to MAP kinases. *Cell Res* 16: 681–692.
3. Tsiftsoglou AS, Tsamadou AI, Papadopoulou LC. 2006. Heme as key regulator of major mammalian cellular functions: Molecular, cellular, and pharmacological aspects. *Pharmacol Ther* 111: 327–345.
4. Hou S, Reynolds MF, Horrigan FT, Heinemann SH, Hoshi T. 2006. Reversible binding of heme to proteins in cellular signal transduction. *Acc Chem Res* 39: 918–924.
5. Padmanaban G, Venkateswar V, Rangarajan PN. 1989. Haem as a multifunctional regulator. *Trends Biochem Sci* 14: 492–496.
6. Zhang L, Hach A. 1999. Molecular mechanism of heme signaling in yeast: The transcriptional activator Hap1 serves as the key mediator. *Cell Mol Life Sci* 56: 415–426.
7. Igarashi K, Hoshino H, Muto A, Suwabe N, Nishikawa S, Nakauchi H, Yamamoto M. 1998. Multivalent DNA binding complex generated by small Maf and Bach1 as a possible biochemical basis for beta-globin locus control region complex. *J Biol Chem* 273: 11783–11790.

8. Oyake T, Itoh K, Motohashi H, Hayashi N, Hoshino H, Nishizawa M, Yamamoto M, Igarashi K. 1996. Bach proteins belong to a novel family of BTB-basic leucine zipper transcription factors that interact with MafK and regulate transcription through the NF-E2 site. *Mol Cell Biol* 16: 6083–6095.

9. Ogawa K, Sun J, Taketani S, Nakajima O, Nishitani C, Sassa S, Hayashi N, Yamamoto M, Shibahara S, Fujita H, Igarashi K. 2001. Heme mediates derepression of Maf recognition element through direct binding to transcription repressor Bach1. *EMBO J* 20: 2835–2843.

10. Sun J, Hoshino H, Takaku K, Nakajima O, Muto A, Suzuki H, Tashiro S, Takahashi S, Shibahara S, Alam J, Taketo MM, Yamamoto M, Igarashi K. 2002. Hemoprotein Bach1 regulates enhancer availability of heme oxygenase-1 gene. *EMBO J* 21: 5216–5224.

11. Wu N, Yin L, Hanniman EA, Joshi S, Lazar MA. 2009. Negative feedback maintenance of heme homeostasis by its receptor, Rev-erbalpha. *Genes Dev* 23: 2201–2209.

12. Yin L, Wu N, Curtin JC, Qatanani M, Szwergold NR, Reid RA, Waitt GM, Parks DJ, Pearce KH, Wisely GB, Lazar MA. 2007. Rev-erbalpha, a heme sensor that coordinates metabolic and circadian pathways. *Science* 318: 1786–1789.

13. Raghuram S, Stayrook KR, Huang P, Rogers PM, Nosie AK, McClure DB, Burris LL, Khorasanizadeh S, Burris TP, Rastinejad F. 2007. Identification of heme as the ligand for the orphan nuclear receptors REV-ERBalpha and REV-ERBbeta. *Nat Struct Mol Biol* 14: 1207–1213.

14. Labbe-Bois R, Labbe P. 1990. Tetrapyrrole and heme biosynthesis in the yeast Sacchromyces cerevisiae. In *Biosynthesis of Heme and Cholorophylls*, ed. HA Dailey, pp. 235–285. New York: Green Pub. Associates and Wiley-Interscience.

15. Mattoon J, Lancashire W, Sanders H, Carvajal E, Malamud D, Braz G, Panek A. 1979. Oxygen and catabolite regulation of hemoprotein biosynthesis in the yeast *Saccharomyces cerevisiae*. In *Biosynthesis of Heme and Cholorophylls*, ed. WJ Caughey, pp. 421–435. New York: Academic Press.

16. Anderson KE, Sassa S, Bishop DF, Desnick RJ. 2009. Disorders of heme biosynthesis: X-linked sideroblastic anemia and the porphyrias. In *The Metabolic and Molecular Bases of Inherited Disease*, eds. CR Scriver, AL Beaudt, WS Sly, D Valle, C Barton, KW Kinzler, B Vogelstein, Chapter 124, pp. 1–53. New York: The McGraw-Hill Companies, Inc.

17. Hon T, Dodd A, Dirmeier R, Gorman N, Sinclair PR, Zhang L, Poyton RO. 2003. A mechanism of oxygen sensing in yeast: Multiple oxygen-responsive steps in the heme biosynthetic pathway affect Hap1 activity. *J Biol Chem* 278: 50771–50780.

18. Pfeifer K, Kim KS, Kogan S, Guarente L. 1989. Functional dissection and sequence of yeast HAP1 activator. *Cell* 56: 291–301.

19. Pfeifer K, Arcangioli B, Guarente L. 1987. Yeast HAP1 activator competes with the factor RC2 for binding to the upstream activation site UAS1 of the CYC1 gene. *Cell* 49: 9–18.

20. Prezant T, Pfeifer K, Guarente L. 1987. Organization of the regulatory region of the yeast CYC7 gene: Multiple factors are involved in regulation. *Mol Cell Biol* 7: 3252–3259.

21. Schneider JC, Guarente L. 1991. Regulation of the yeast CYT1 gene encoding cytochrome c1 by HAP1 and HAP2/3/4. *Mol Cell Biol* 11: 4934–4942.

22. Verdiere J, Creusot F, Guarente L, Slonimski PP. 1986. The overproducing CYP1 and the underproducing hap1 mutations are alleles of the same gene which regulates in trans the expression of the structural genes encoding iso-cytochromes c. *Curr Genet* 10: 339–342.

23. Zitomer RS, Lowry CV. 1992. Regulation of gene expression by oxygen in Saccharomyces cerevisiae. *Microbiol Rev* 56: 1–11.

24. Bock KW, De Matteis F, Aldridge WN. 1978. *Heme and Hemoproteins.* New York: Springer-Verlag.

25. Johnston M. 1987. A model fungal gene regulatory mechanism: The GAL genes of Saccharomyces cerevisiae. *Microbiol Rev* 51: 458–476.

26. Zhang L, Bermingham MO, Turcotte B, Guarente L. 1993. Antibody-promoted dimerization bypasses the regulation of DNA binding by the heme domain of the yeast transcriptional activator HAP1. *Proc Natl Acad Sci U S A* 90: 2851–2855.

27. Marmorstein R, Carey M, Ptashne M, Harrison SC. 1992. DNA recognition by GAL4: Structure of a protein–DNA complex [see comments]. *Nature* 356: 408–414.

28. Marmorstein R, Harrison SC. 1994. Crystal structure of a PPR1-DNA complex: DNA recognition by proteins containing a Zn2Cys6 binuclear cluster. *Genes Dev* 8: 2504–2512.

29. Hach A, Hon T, Zhang L. 2000. The coiled coil dimerization element of the yeast transcriptional activator Hap1, a Gal4 family member, is dispensable for DNA binding but differentially affects transcriptional activation. *J Biol Chem* 275: 248–254.

30. Reece RJ, Ptashne M. 1993. Determinants of binding-site specificity among yeast C6 zinc cluster proteins. *Science* 261: 909–911.

31. Zhang L, Guarente L. 1994. The yeast activator HAP1 — a GAL4 family member — binds DNA in a directly repeated orientation. *Genes Dev* 8: 2110–2119.

32. Zhang L, Guarente L. 1996. The C6 zinc cluster dictates asymmetric binding by HAP1. *EMBO J* 15: 4676–4681.

33. King DA, Zhang L, Guarente L, Marmorstein R. 1999. Structure of a HAP1/DNA complex reveals dramatically asymmetric DNA binding by a homodimeric protein. *Nat Struct Biol* 6: 64–71.

34. Hach A, Hon T, Zhang L. 1999. A new class of repression modules is critical for heme regulation of the yeast transcriptional activator Hap1. *Mol Cell Biol* 19: 4324–4333.

35. Hon T, Hach A, Lee HC, Cheng T, Zhang L. 2000. Functional analysis of heme regulatory elements of the transcriptional activator Hap1. *Biochem Biophys Res Commun* 273: 584–591.

36. Kitanishi K, Igarashi J, Hayasaka K, Hikage N, Saiful I, Yamauchi S, Uchida T, Ishimori K, Shimizu T. 2008. Heme-binding characteristics of the isolated PAS-A domain of mouse Per2, a transcriptional regulatory factor associated with circadian rhythms. *Biochemistry* 47: 6157–6168.

37. Zhang L, Hach A, Wang C. 1998. Molecular mechanism governing heme signaling in yeast: A higher-order complex mediates heme regulation of the transcriptional activator HAP1. *Mol Cell Biol* 18: 3819–3828.

38. Hon T, Lee HC, Hach A, Johnson JL, Craig EA, Erdjument-Bromage H, Tempst P, Zhang L. 2001. The Hsp70-Ydj1 molecular chaperone represses the activity of the heme activator protein Hap1 in the absence of heme. *Mol Cell Biol* 21: 7923–7932.

39. Lee HC, Hon T, Zhang L. 2002. The molecular chaperone Hsp90 mediates heme activation of the yeast transcriptional activator Hap1. *J Biol Chem* 277: 7430–7437.

40. Lan C, Lee HC, Tang S, Zhang L. 2004. A novel mode of chaperone action: heme activation of Hap1 by enhanced association of Hsp90 with the repressed Hsp70-Hap1 complex. *J Biol Chem* 279: 27607–27612.

41. Picard D. 1998. The role of heat-shock protein in the regulation of steroid receptor function. In *Molecular Biology of Steroid and Nuclear Hormone Receptors*, ed. LP Freedman, pp. 1–18. Boston: Birkhauser.

42. Lee HC, Zhang L. 2009. A unique mechanism of chaperone action: Heme regulation of Hap1 activity involves separate control of repression and activation. *Protein Pept Lett* 16: 642–649.

43. Pratt W. 1997. The role of the hsp90-based chaperone system in signal transduction by nuclear receptors and receptors signaling via MAP kinase. *Annu Rev Pharmacol Toxicol* 37: 297–326.

44. Pratt W, Toft D. 1997. Steroid receptor interactions with heat shock protein and immunophilin chaperones. *Endocr Rev* 18: 306–360.

45. Pratt WB. 1998. The hsp90-based chaperone system: involvement in signal transduction from a variety of hormone and growth factor receptors. *Proc Soc Exp Biol Med* 217: 420–434.

46. Pratt WB, Dittmar KD. 1998. Studies with purified chaperones advance the understanding of the mechanism of glucocorticoid receptor-hsp90 heterocomplex assembly. *Trends Endocrinol Metab* 9: 244–252.

47. Pratt WB, Toft DO. 2003. Regulation of signaling protein function and trafficking by the Hsp90/Hsp70-based chaperone machinery. *Exp Biol Med (Maywood)* 228: 111–133.

48. Johnson J, Corbisier R, Stensgard B, Toft D. 1996. The involvement of p23, Hsp90, and immunophilins in the assembly of progesterone receptor complexes. *Steroid Biochem Mol Biol* 56: 31–37.

49. Yamashita S. 1998. Localization and functions of steroid hormone receptors. *Histol Histopathol* 13: 255–270.

50. Duina A, Chang H, Marsh J, Lindquist S, Gaber R. 1996. A cyclophilin function in Hsp90-dependent signal transduction [see comments]. *Science* 274: 1713–1715.

51. Duina AA, Kalton HM, Gaber RF. 1998. Requirement for Hsp90 and a CyP-40-type cyclophilin in negative regulation of the heat shock response. *J Biol Chem* 273: 18974–18978.

52. Xu Z, Pal JK, Thulasiraman V, Hahn HP, Chen JJ, Matts RL. 1997. The role of the 90-kDa heat-shock protein and its associated cohorts in stabilizing the heme-regulated eIF-2alpha kinase in reticulocyte lysates during heat stress. *Eur J Biochem* 246: 461–470.

53. Dittmar K, Banach M, Galigniana M, Pratt W. 1998. The role of DnaJ-like proteins in glucocorticoid receptor.hsp90 heterocomplex assembly by the reconstituted hsp90.p60.hsp70 foldosome complex. *J Biol Chem* 273: 7358–7366.

54. Dittmar K, Demady D, Stancato L, Krishna P, Pratt W. 1997. Folding of the glucocorticoid receptor by the heat shock protein (hsp) 90-based chaperone machinery. The role of p23 is to stabilize receptor.hsp90 heterocomplexes formed by hsp90.p60.hsp70. *J Biol Chem* 272: 21213–21220.

55. Smith DF, Faber LE, Toft DO. 1990. Purification of unactivated progesterone receptor and identification of novel receptor-associated proteins. *J Biol Chem* 265: 3996–4003.

56. Schumacher RJ, Hansen WJ, Freeman BC, Alnemri E, Litwack G, Toft DO. 1996. Cooperative action of Hsp70, Hsp90, and DnaJ proteins in protein renaturation. *Biochemistry* 35: 14889–14898.

57. Toft DO. 1998. Recent advances in the study of Hsp90 structure and mechanism of action. *Trends Endocrinol Metab* 9: 238–243.

58. Lee HC, Hon T, Lan C, Zhang L. 2003. Structural environment dictates the biological significance of heme-responsive motifs and the role of Hsp90 in the activation of the heme activator protein Hap1. *Mol Cell Biol* 23: 5857–5866.

59. Cyr DM, Lu X, Douglas MG. 1992. Regulation of Hsp70 function by a eukaryotic DnaJ homolog. *J Biol Chem* 267: 20927–20931.

60. Caplan AJ, Langley E, Wilson EM, Vidal J. 1995. Hormone-dependent transactivation by the human androgen receptor is regulated by a DnaJ protein. *J Biol Chem* 270: 5251–5257.

61. Kagami M, Toh-e A, Matsui Y. 1997. SRO9, a multicopy suppressor of the bud growth defect in the *Saccharomyces cerevisiae* rho3-deficient cells, shows strong genetic interactions with tropomyosin genes, suggesting its role in organization of the actin cytoskeleton. *Genetics* 147: 1003–1016.

62. Hoshino H, Igarashi K. 2002. Expression of the oxidative stress-regulated transcription factor bach2 in differentiating neuronal cells. *J Biochem (Tokyo)* 132: 427–431.

63. Kitamuro T, Takahashi K, Ogawa K, Udono-Fujimori R, Takeda K, Furuyama K, Nakayama M, Sun J, Fujita H, Hida W, Hattori T, Shirato K, Igarashi K, Shibahara S. 2003. Bach1 functions as a hypoxia-inducible repressor for the heme oxygenase-1 gene in human cells. *J Biol Chem* 278: 9125–9133.

64. Ishikawa M, Numazawa S, Yoshida T. 2005. Redox regulation of the transcriptional repressor Bach1. *Free Radic Biol Med* 38: 1344–1352.

65. MacLeod AK, McMahon M, Plummer SM, Higgins LG, Penning TM, Igarashi K, Hayes JD. 2009. Characterization of the cancer chemopreventive NRF2-dependent gene battery in human keratinocytes: Demonstration that the KEAP1-NRF2 pathway, and not the BACH1-NRF2 pathway, controls cytoprotection against electrophiles as well as redox-cycling compounds. *Carcinogenesis* 30: 1571–1580.

66. Zhong JL, Raval C, Edwards GP, Tyrrell RM. 2009. A role for Bach1 and HO-2 in suppression of basal and UVA-induced HO-1 expression in human keratinocytes. *Free Radic Biol Med* 48: 196–206.

67. Tanimoto T, Hattori N, Senoo T, Furonaka M, Ishikawa N, Fujitaka K, Haruta Y, Yokoyama A, Igarashi K, Kohno N. 2009. Genetic ablation of the Bach1 gene reduces hyperoxic lung injury in mice: Role of IL-6. *Free Radic Biol Med* 46: 1119–1126.

68. Kassovska-Bratinova S, Yang G, Igarashi K, Dennery PA. 2009. Bach1 modulates heme oxygenase-1 expression in the neonatal mouse lung. *Pediatr Res* 65: 145–149.

69. Peterson KR. 2003. Hemoglobin switching: New insights. *Curr Opin Hematol* 10: 123–129

70. Johnson RA, Johnson FK. 2000. The effects of carbon monoxide as a neurotransmitter. *Curr Opin Neurol* 13: 709–713.

71. Mancuso C, Perluigi M, Cini C, De Marco C, Giuffrida Stella AM, Calabrese V. 2006. Heme oxygenase and cyclooxygenase in the central nervous system: a functional interplay. *J Neurosci Res* 84: 1385–1391.

72. Navarra P, Dello Russo C, Mancuso C, Preziosi P, Grossman A. 2000. Gaseous neuromodulators in the control of neuroendocrine stress axis. *Ann NY Acad Sci* 917: 638–646.

73. Tomaro ML, Batlle AM. 2002. Bilirubin: Its role in cytoprotection against oxidative stress. *Int J Biochem Cell Biol* 34: 216–220.

74. Sun J, Muto A, Hoshino H, Kobayashi A, Nishimura S, Yamamoto M, Hayashi N, Ito E, Igarashi K. 2001. The promoter of mouse transcription repressor bach1 is regulated by Sp1 and trans-activated by Bach1. *J Biochem (Tokyo)* 130: 385–392.

75. Suzuki H, Tashiro S, Hira S, Sun J, Yamazaki C, Zenke Y, Ikeda-Saito M, Yoshida M, Igarashi K. 2004. Heme regulates gene expression by triggering Crm1-dependent nuclear export of Bach1. *EMBO J* 23: 2544–2553.

76. Suzuki H, Tashiro S, Sun J, Doi H, Satomi S, Igarashi K. 2003. Cadmium induces nuclear export of Bach1, a transcriptional repressor of heme oxygenase-1 gene. *J Biol Chem* 278: 49246–49253.

77. Hira S, Tomita T, Matsui T, Igarashi K, Ikeda-Saito M. 2007. Bach1, a heme-dependent transcription factor, reveals presence of multiple heme binding sites with distinct coordination structure. *IUBMB Life* 59: 542–551.

78. Zhang L, Guarente L. 1995. Heme binds to a short sequence that serves a regulatory function in diverse proteins. *EMBO J* 14: 313–320.

79. Zenke-Kawasaki Y, Dohi Y, Katoh Y, Ikura T, Ikura M, Asahara T, Tokunaga F, Iwai K, Igarashi K. 2007. Heme induces ubiquitination and degradation of the transcription factor Bach1. *Mol Cell Biol* 27: 6962–6971.

80. Goodsell DS. 2003. The molecular perspective: Ubiquitin and the proteosome. *Oncologist* 8: 293–294.

81. Ben-Neriah Y. 2002. Regulatory functions of ubiquitination in the immune system. *Nat Immunol* 3: 20–26.

82. Ande SR, Chen J, Maddika S. 2009. The ubiquitin pathway: An emerging drug target in cancer therapy. *Eur J Pharmacol* 625: 199–205.

83. Herrmann J, Lerman LO, Lerman A. 2007. Ubiquitin and ubiquitin-like proteins in protein regulation. *Circ Res* 100: 1276–1291.

84. Taylor C, Jobin C. 2005. Ubiquitin protein modification and signal transduction: Implications for inflammatory bowel diseases. *Inflamm Bowel Dis* 11: 1097–1107.

85. Laney JD, Hochstrasser M. 1999. Substrate targeting in the ubiquitin system. *Cell* 97: 427–430.

86. Bayle J, Lopez S, Iwai K, Dubreuil P, De Sepulveda P. 2006. The E3 ubiquitin ligase HOIL-1 induces the polyubiquitination and degradation of SOCS6 associated proteins. *FEBS Lett* 580: 2609–2614.

87. Ishikawa H, Kato M, Hori H, Ishimori K, Kirisako T, Tokunaga F, Iwai K. 2005. Involvement of heme regulatory motif in heme-mediated ubiquitination and degradation of IRP2. *Mol Cell* 19: 171–181.

88. Wang H, Zhou Q, Kesinger JW, Norris C, Valdez C. 2007. Heme regulates exocrine peptidase precursor genes in zebrafish. *Exp Biol Med (Maywood)* 232: 1170–1180.

Chapter **4**

HEME-REGULATED eIF2α KINASE IN TRANSLATION AND ERYTHROPOIESIS

Jane-Jane Chen and Rajasekhar NVS Suragani

Overview

Besides serving as a prosthetic group for hemoglobin, heme also regulates translation in erythroid precursors by modulating the activity of heme-regulated eIF2α kinase (HRI). HRI balances synthesis of heme and globin by sensing the intracellular heme concentration through its two heme-binding domains. In heme deficiency, HRI is activated by autophosphorylation, and phosphorylates the α-subunit of eukaryotic initiation factor (eIF2α), which impairs the recycling of eIF2 for translational initiation and results in cessation of protein synthesis. In the absence of HRI, uncontrolled protein synthesis in heme deficiency results in globin aggregation and precipitation in the red blood cell and its precursors. In addition to general inhibition of protein synthesis, phosphorylation of eIF2α by HRI in erythroid precursors also leads to a selective increase in the translation of ATF4 mRNA. The induction of gene transcription by ATF4 may help cells mitigate and adapt to stress. Indeed, HRI is necessary for adaptive gene expression in erythroid precursors under the stress of iron deficiency. In the absence of HRI, stress erythropoiesis during iron deficiency is hampered by a severe inhibition of erythroid differentiation. Thus, translational control by HRI not only regulates hemoglobin synthesis but also erythroid differentiation during stress erythropoiesis.

55

Introduction

Iron deficiency anemia is very common with an incidence of approximately 1 billion cases worldwide and is marked clinically by microcytic hypochromic red blood cells (reviewed in Refs. 1–2). Microcytosis and hypochromia in iron deficiency has led to earlier studies, which demonstrate the stimulation of hemoglobin synthesis by inorganic iron in immature erythroid cells (3–5). Further studies showed that heme, not iron *per se*, is required for protein synthesis in reticulocyte lysates since the iron-chelating agent, desferrioxamine, does not block the stimulatory effect of heme (6, 7). Thus, in addition to serving as a prosthetic group for hemoglobin, heme also regulates translation in erythroid precursors. Iron is incorporated into protoporphyrin IX in the last step of heme biosynthesis by ferrochelatase to form heme. Consequently, iron deficiency leads to heme deficiency. Heme-iron accounts for the majority of the iron in the human body, and hemoglobin contains as much as 70% of the total iron content of a normal adult. Accordingly, iron and heme homeostasis plays very important roles in hemoglobin synthesis and erythroid differentiation.

Mature red blood cells (RBCs) are packed with a very high concentration of hemoglobin, approximately 5 mM (340 mg/mL) in both humans and mice. It is critical during erythroid differentiation and maturation that the three components of hemoglobin, α-globin, β-globin and heme, are made in the 2:2:4 ratio in order to form stable $\alpha_2\beta_2$ hemoglobin complexed with four heme molecules. Imbalance of these three components can be deleterious since each component in excess is cytotoxic to RBCs and their precursors. The significance of the globin to heme ratio is well demonstrated by the hypochromic anemia in iron deficiency. Translational control by the heme-regulated inhibitor kinase (HRI) provides one major mechanism to ensure the balanced synthesis of globins and heme. Additionally, heme also regulates transcription of globin and heme oxygenase-1 genes (8). In this review, we will focus on the roles of heme and HRI in translation and erythropoiesis.

HRI is responsible for heme-regulated translation in erythroid precursors

Protein synthesis in erythroid precursors is dependent on the availability of heme. Heme deficiency causes disaggregation of polysomes and inhibition

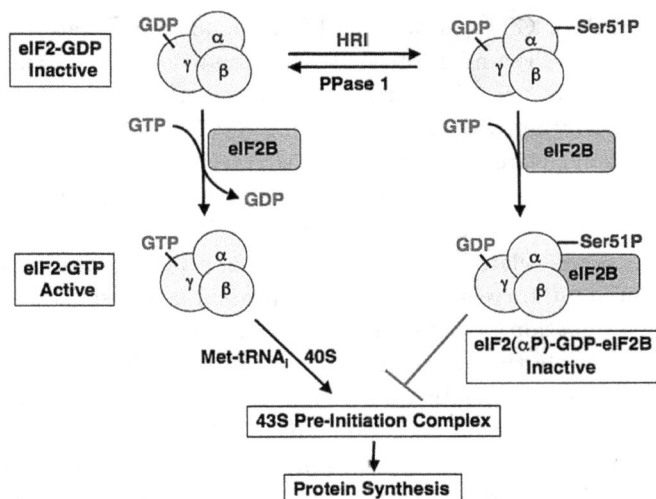

Fig. 1.   Inhibition of protein synthesis by phosphorylation of eIF2α. This figure was originally published in Blood. Chen J-J, Regulation of protein synthesis by heme-regulated eIF2α kinase: Relevance to anemias. *Blood* 2007, **109**: 2693–2699 © The American Society of Hematology.

of protein synthesis in reticulocytes (7, 9, 10). On addition of hemin, polysomes reform and protein synthesis is restored (7, 11). This inhibition of protein synthesis in heme deficiency is associated with a marked decrease in the formation of 40S·eIF2·Me$_{\overline{\imath}}$-tRNA$_i$GTP (the 43S pre-initiation complex) (12), and with a rapid increase in the phosphorylation of the α subunit of eIF2 (13, 14). The phosphorylation of eIF2α is concomitant with the activation of HRI (15–17). HRI was later shown to be the heme-regulated eIF2α kinase that phosphorylates the α-subunit of eIF2 [(18, 19) and reviewed in (20)].

The molecular mechanism of inhibition of translational initiation by the phosphorylation of eIF2α has been studied extensively [reviewed in (21, 22)], and is illustrated in Fig. 1. eIF2 is a heterotrimeric protein which binds GTP, initiating Met-tRNA$_i$ and the 40S ribosomal subunit to form the 43S pre-initiation complex. eIF2 exists in two forms, the inactive eIF2-GDP and the active eIF2-GTP. The GTP in the eIF2-GTP complex is hydrolyzed to GDP upon binding of the 60S ribosomal subunit to the 43S pre-initiation complex during translation. The recycling of eIF2 for another round of translational initiation therefore requires the exchange of its bound GDP for GTP. However, eIF2 has a 400-fold greater affinity for GDP than for GTP under physiological conditions. This exchange of tightly bound GDP for GTP requires another initiation factor eIF2B, which is rate limiting with a concentration of 15% to 25% of eIF2. When eIF2α is phosphorylated by HRI at Ser51, the phosphorylated eIF2(αP)-GDP binds much more tightly to eIF2B than eIF2-GDP does and prevents the GDP/GTP exchange activity of eIF2B (23). Thus, once the amount of phosphorylated eIF2 exceeds that of eIF2B, protein synthesis is shut off. The inhibitory effect of HRI on protein synthesis is reversed following heme repletion [reviewed in (20)]. This recovery of protein synthesis requires not only the inactivation of HRI by heme but also the removal of the inhibitory phosphate at Ser51 of eIF2α by type-1 phosphatase (PPase 1) (24–29) to regenerate the active eIF2-GTP (Fig. 1).

Studies using polyclonal and monoclonal antibodies to HRI have shown that HRI is principally responsible for the inhibition of protein synthesis during heme deficiency (30, 31). Using targeted disruption of the HRI gene in mice, Han *et al.* have demonstrated that HRI-null ($Hri^{-/-}$) reticulocytes have an increase in protein synthesis along with significantly less eIF2α phosphorylation as compared to $Hri^{+/+}$ reticulocytes (32). This increase in protein synthesis is accompanied by an increase in larger size polysomes in $Hri^{-/-}$ reticulocytes. There is no difference in the β-globin mRNA levels between Hri +/+ and −/− reticulocytes. Therefore, $Hri^{-/-}$ reticulocytes have a higher rate of translation initiation, providing *in vivo* evidence for the role of eIF2α phosphorylation by HRI in the regulation of translational initiation. Furthermore, protein synthesis in $Hri^{-/-}$ reticulocytes does not increase with the addition of hemin as seen in $Hri^{+/+}$ reticulocytes. These results establish unequivocally that HRI is essential for heme-regulated translation in reticulocytes (32).

Phosphorylation of eIF2α under diverse cellular stresses by the family of eIF2α kinases

Phosphorylation of eIF2α occurs under diverse stress conditions in addition to heme deficiency, and is a key regulatory step in regulation of protein synthesis that is conserved in eukaryotes from yeast to human [reviewed in (33)]. Indeed, preemptive phosphorylation of eIF2α protects cells from oxidative and endoplasmic reticulum (ER) stresses (34). Besides HRI, three other eIF2α kinases are known: the double-stranded RNA-dependent eIF2α kinase (PKR), the GCN2 protein kinase and the

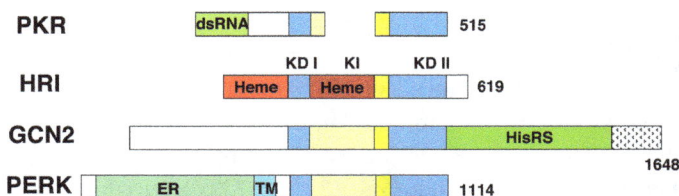

Fig. 2. The family of eIF2α kinases: protein domains, regulators and tissue expression. All four mammalian eIF2α kinases share amino acid sequence homology in their kinase domains (shaded blue). The signature sequence of eIF2α kinases preceded conserved kinase domain 2 is highlighted in yellow. Unique domains of each kinase are also labeled. The total number of amino acid residues in each kinase from mice is shown at the end of each kinase schematic. KD1: kinase domain 1; KD2: kinase domain 2; HisRS, histidyle tRNA synthase-like domain; ER, ER sensing domain; TM, transmembrane domain.

PKR-like ER resident kinase (PERK). This family of eIF2α kinases share extensive homology in their kinase catalytic domains (35–41) and phosphorylate eIF2α at the same Ser51 residue (40–43). While they share a common mode of action, each kinase elicits a different physiological response as a consequence of both distinct tissue distributions and different signals to which they respond via their unique regulatory domains (Fig. 2).

PKR responds to viral infection [reviewed in (44)] while GCN2 senses nutrient starvations [reviewed in (45)]. PERK is activated by ER stress [reviewed in (46)], and HRI is regulated by heme [reviewed in (20)]. PKR is expressed ubiquitously and is induced by interferon. GCN2 is highly expressed in the brain and liver while PERK is highly expressed in the pancreas. HRI is highly expressed in erythroid precursors and is induced during erythroid differentiation (47, 48). The unique function of each of these eIF2α kinases in response to different stresses has been validated by phenotypes of knockout mice deficient in each of four eIF2α kinases.

Pkr$^{-/-}$ mice are compromised in their ability to respond to viral challenges (49, 50), and *Perk*$^{-/-}$ mice develop diabetes between 2 and 4 weeks of age (51). *Gcn2*$^{-/-}$ mice have reduced viability upon amino acid starvation (52) and lack the ability to discriminate imbalanced diets (53, 54). GCN2 is also involved in memory acquisition (55). The erythroid response of *Hri*$^{-/-}$ mice to heme deficiency is abnormal (32, 56, 57). HRI is also necessary to reduce phenotypic severities of inherited red cell disorders of β-thalassemia and erythroprotoporphyria (56). The knockin mice homozygous for the nonphosphorylatable eIF2α Ser51Ala mutation (*eIF2α A/A mice*) at the endogenous eIF2α loci cannot be phosphorylated by eIF2α kinases; therefore they are defective in the regulation by eIF2αP. *eIF2α A/A* mice have more severe phenotypes than knockout mice of each eIF2α kinase and die of hypoglycemia shortly after birth (58).

Activation of HRI by autophosphorylation

In heme-deficient reticulocyte lysates, inhibition of protein synthesis is accompanied by the increased phosphorylation of both HRI and eIF2α (13). HRI is present as a homodimer and no significant change in the molecular size of HRI is observed upon its activation in heme deficiency (59–61). Using inactive HRI mutants in which the conserved K199 in rabbit HRI and K196 in mouse HRI have been changed to Arg, we have established unequivocally that HRI is activated by autophosphorylation (64, 65, 67, 68). Purified HRI undergoes intermolecular multiple autophosphorylation in the absence of heme (62–65). HRI autophosphorylation correlates well with the conversion of inactive ProHRI to activated HRI in heme deficiency (62). There are eight major tryptic phosphopeptides in fully autophosphorylated HRI (66).

The model depicted in Fig. 3 summarizes findings from our laboratory on the activation of HRI by multiple autophosphorylation in heme deficiency. We propose that newly synthesized HRI is rapidly dimerized and undergoes multiple intermolecular autophosphorylation in heme deficiency in three stages. In the first stage, autophosphorylation of newly synthesized HRI stabilizes the ProHRI against aggregation. Although ProHRI is an active autokinase, it lacks eIF2α kinase activity. Additional multiple autophosphorylation in the second stage is required for the formation of stable dimeric heme-reversible HRI. In heme abundance, heme binds to the second heme-binding site in the kinase insert (KI) domain and represses the activation of heme-reversible HRI. In the heme deficiency, heme-reversible HRI undergoes the final third stage of autophosphorylation,

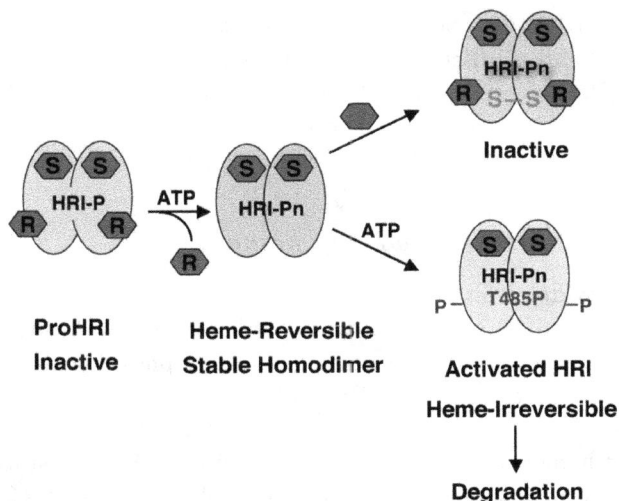

Fig. 3. A schematic model of the activation of HRI by multiple auto-phosphorylation. During the biosynthesis of HRI in hemin-supplemented reticulocyte lysates, Hsp90 and Hsc70 are associated co-translationally with HRI (117, 118). These chaperones are necessary for proper folding and maintenance of HRI in its inactive ProHRI form, which can be activated by heme deficiency. It is most likely that both heme-binding sites of the ProHRI have heme bound. In heme deficiency, ProHRI is converted to stable heme-reversible HRI by dissociation of the heme bound to the kinase insert (KI) domain and subsequent autophosphorylation (74). Purified heme-reversible HRI is a stable homodimer held together by non-covalent interactions. This HRI is autophosphorylated and has one heme stably bound per subunit to its N-terminal domain, and is an active autokinase regulated by heme (64). Further autophosphorylation of HRI at Thr485 activates its eIF2α kinase activity (66). This activated HRI is no longer regulated by heme (heme-irreversible HRI) and is degraded (66). In heme abundance, heme binds to the KI domain of HRI, promotes intersubunit disulfide formation in the HRI homodimer and inhibits kinase activity. Heme molecules are represented by hexagons; S for stable heme-binding while R for reversible heme-binding.

attains eIF2α kinase activity and is no longer regulated by heme (heme-irreversible HRI).

Thr485 in the activation loop of mouse HRI is one of the residues autophosphorylated by heme-reversible HRI and is essential for attaining eIF2α kinase activity in heme deficiency as well as in sodium arsenite-induced activation of HRI (66). Several other potential autophosphorylation sites at the second and third stages of HRI autophosphorylation have also been identified (Rafie-Kolpin and Chen unpublished). The functions of the phosphorylation of each of these residues in the activation, stability and solubility of HRI remain to be investigated.

Inhibition of HRI activity by heme

Hemin (the oxidized form of heme with Fe^{+3}) has been shown to bind to highly purified HRI. This binding results in the inhibition of both autophosphorylation and HRI-mediated eIF2α phosphorylation (63, 64, 69). Native rabbit reticulocyte HRI and recombinant HRI expressed from baculovirus-infected insect Sf9 cells (BV-HRI) and E. coli have been purified to near homogeneity (64, 65). This purified HRI is a stable homodimer, undergoes autophosphorylation and phosphorylates eIF2α. Both autokinase and eIF2α kinase activities are inhibited by submicromolar concentrations of hemin with an apparent K_i (concentration at 50% inhibition) of 0.2 µM. The ATP binding to HRI is inhibited by prior treatment of HRI with hemin in a concentration-dependent manner (70). Thus, inhibition of autophosphorylation and eIF2α phosphorylation of HRI by heme is likely the result of blocking ATP binding to HRI.

Hemin also promotes intersubunit disulfide bond formation in HRI homodimers (61, 64, 71). There is a positive correlation between intersubunit disulfide formation of HRI and the inhibition of HRI kinase activities *in vitro* by various porphyrin compounds that are structurally similar to hemin. Furthermore, there is a positive correlation between the abilities of various porphyrin compounds to promote intersubunit disulfide formation of HRI and the maintenance of protein synthesis, the reversal of the inhibition of protein synthesis and the phosphorylation of eIF2α in rabbit reticulocyte lysates (61). These findings underscore the importance of free sulfhydryls and intersubunit disulfide bond formation in the regulation of HRI by heme (Fig. 3).

Two distinct heme-binding domains in the cooperation of heme-regulation of HRI

Earlier study of Chefalo *et al.* has shown that BV-HRI purifies as a hemoprotein with the three characteristic absorption peaks in the visible wavelength, the Soret band at 424 nm and the α and β bands around 550 nm (64). Sf 9 cells are low in heme content, and expression of some functional

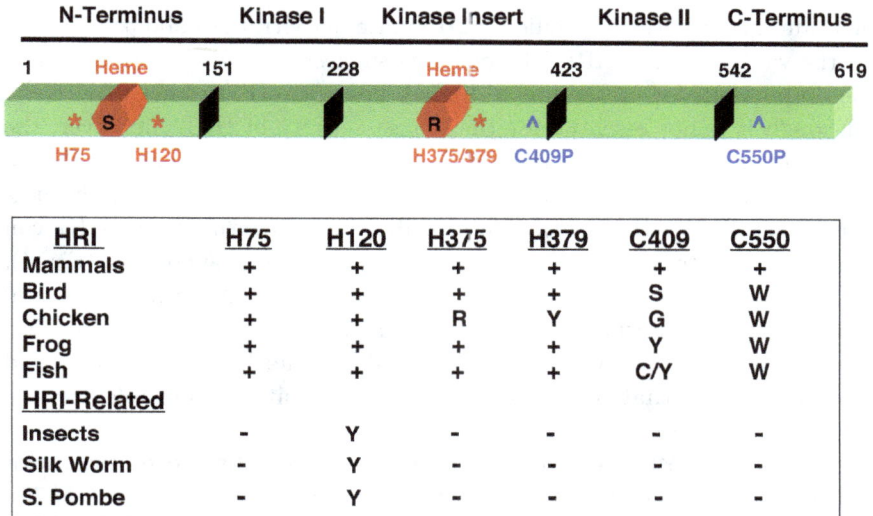

| N-Terminus | | Kinase I | | Kinase Insert | | Kinase II | | C-Terminus | |

HRI	H75	H120	H375	H379	C409	C550
Mammals	+	+	+	+	+	+
Bird	+	+	+	+	S	W
Chicken	+	+	R	Y	G	W
Frog	+	+	+	+	Y	W
Fish	+	+	+	+	C/Y	W
HRI-Related						
Insects	-	Y	-	-	-	-
Silk Worm	-	Y	-	-	-	-
S. Pombe	-	Y	-	-	-	-

Fig. 4. Protein domains of HRI and the evolutionary conservation of heme coordination in HRI. HRI is divided into five domains as indicated. The amino acid sequence of mouse HRI is used here. Heme molecules are marked in red; S denotes the stable heme-binding site while R denotes the reversible heme-binding site. * marks the histidine residues that coordinate the heme molecule. ^ marks the positions of CP, putative heme-regulated motifs.

hemoproteins such as cytochrome P450 and nitric oxide synthase (72, 73) requires the addition of hemin to the culture medium. Thus, the finding that HRI purifies as a hemoprotein from Sf 9 cells without added exogenous heme indicates that HRI has a high affinity for heme *in vivo*. Furthermore, there are two distinct types of heme-binding sites per HRI homodimer (64). One type of binding site is nearly saturated with stably-bound endogenous heme co-purified with HRI while the other binding site is available to bind hemin reversibly (Fig. 4). This second reversible heme-binding site is likely to be responsible for the down-regulation of HRI activity by heme (64). The stoichiometry of two heme molecules per HRI monomer has been confirmed by direct measurement of heme chromophore through alkaline pyridine treatment of homogeneous heme-saturated HRI (65).

As shown above in Fig. 2, there are three unique regions of HRI in the N-terminus, the KI and the C-terminus. The N-terminal 130 amino acids of HRI are necessary for stable high-affinity heme-binding to HRI. N-terminally truncated Met2 (Δ103) and Met3 (Δ130) HRI are active eIF2α kinases and autokinases; their specific eIF2α kinase activities are about 50% that of the wild type (wt) HRI. These results suggest that the N-terminus may be important for achieving a higher specific eIF2α kinase activity, although it is not essential for the kinase activity of HRI. Moreover, the N-terminus is essential for the high sensitivity of HRI to heme regulation since Met2 and Met3 HRI are 10 times less sensitive to heme inhibition as compared to Wt HRI (74). Additionally, both N-terminus and KI can bind heme whereas Kinase I, Kinase II, and C-terminus cannot (74) (Fig. 4). When expressed in the presence of 5 μM hemin, the N-terminal domain, but not the KI, purified as a hemoprotein with a visible pink color. This observation together with the loss of stable heme binding in Met2 and Met3, indicate that KI is the heme-binding site responsible for the reversible heme-regulation of HRI (74).

Conserved residues His75 and His120 were identified to be the proximal and distal heme ligands respectively in the N-terminus domain [(75, 76), and our unpublished observations] as illustrated in Fig. 4. Mutation of His75 and His120 individually to Ala in full-length HRI resulted in decreased sensitivity to heme inhibition, similar to N-terminally truncated HRI (Yen and Chen unpublished). This finding further underscores the importance of heme coordination in the N-terminus in the down-regulation of HRI activity. Thus, HRI is a unique and novel hemoprotein with dual heme-binding domains. Furthermore, there is cooperation between the two heme-binding domains to achieve efficient heme regulation of HRI kinase activities. In this regard, heme has been shown to stabilize the binding of the N-terminus domain with N-terminally-truncated HRI (77) as well as to stabilize the binding of the N-terminus domain with the KI domain (78).

HRI contains two putative heme regulatory motifs (HRMs) (79–81), Cys409/Pro410 and Cys550/Pro 551 in mouse HRI, which are not present in the other three members of eIF2α kinases (Fig. 4). The significance of these HRM motifs in heme regulation of HRI is not clear. These two Cys residues have been mutated individually to Ser with no apparent effect on the heme responsiveness of HRI (74). Additionally, these two Cys residues are not conserved in chicken, xenopus or fish HRI (82, 83), indicating that they may not be important for the heme regulation of HRI. However, a recent study suggests that His120 and Cys409 are heme ligands in the full-length HRI (78). The proposed role of Cys409 by Igarashi *et al.* in the regulation of HRI by heme is in conflict with the fact that this Cys residue is not conserved in HRI from fish, xenopus, bird and chicken. One possible explanation of the discrepancy between these results is the use of different expression systems to obtain recombinant HRI. In contrast to the *E. coli* system (65), Sf 9 expression system permitted heme incorporation into the N-terminus of HRI (64). Furthermore, HRI expressed in the *E. coli* system is also multiple hyperphosphrylated (65).

Tissue expression and phylogenetic conservation of HRI

HRI protein and mRNA are expressed highly in erythropoietic tissues of the bone marrow, spleen and fetal liver (47). In addition, HRI mRNA levels increase during erythroid differentiation of mouse erythroleukemic (MEL) cells, and this increase is dependent on the presence of heme (47). Similarly, HRI mRNA is increased during erythroid differentiation of G1E-Er4 embryonic stem cells (84). Most significantly, HRI expression is also increased during erythroid differentiation of fetal livers *in vivo* (48). HRI is not detectable in mature RBCs (85). While there are increasing reports of the expression of HRI in non-erythroid cells (39, 86–89), it is important to note that HRI is expressed at two orders of magnitude higher in erythroid precursors as compared to non-erythroid cells such as macrophages (57).

HRI homologs are present in vertebrates with amino acid sequence identity to mouse HRI ranging from 45% in fish to 82% in humans. Importantly, the two heme ligands in the N-terminal heme-binding domain described above, His75 and His120, are conserved in all vertebrate HRI (Fig. 4). Interestingly, *Schizosaccharomyces pombe* not only has GCN2 but also has two additional eIF2α kinases, which have greater homology to HRI (27% identity, and 45% homology) than to other eIF2α kinases. Although these two HRI-related eIF2α kinases can be regulated by heme *in vitro*, albeit with a higher Ki, these HRI-related kinases do not respond to iron/heme deficiency *in vivo* (90) as mammalian HRI does (32). They do, however, respond to oxidative and heat stress (90, 91) like mouse HRI in erythroid precursors (68). Searches of the current NCBI database reveal the presence of the *pombe*-like HRI related kinases in *Anapheles gambiae, Aedes aegypti, Tribolium castaneum, Bombyx mori, Branchiostoma floridae, Ciona intestinalis* and *Nematostella vectensis*. The N-terminus sequences of these HRI-related kinases are much less conserved than those of vertebrate HRI. Furthermore, His75 is no longer conserved while His120 is replaced with Tyr in these HRI-related kinases. There is no HRI in *Saccharomyces cerevisiae, Caenorhabditis elegans* or *Drosophila melanogaster*. Although it has been suggested that the malarial parasite eIF2α kinase, PfPk4, is heme-regulated (92), no significant homology has been found to the heme-binding domains of mammalian HRI.

The higher degree of conservation of vertebrate HRI from fish to human, the high level expression of HRI in erythroid cells and the high sensitivity of HRI to heme regulation support the notion that HRI evolved when diffusion alone could no longer supply the increasing demands for oxygen and blood circulation was established wherein hemoglobin is carried in red blood cells. Vertebrate HRI retains its ability to respond to oxidative stress and heat stress (68) similar to the *pombe* HRI-related kinase (91).

Coordination of heme and globin synthesis by HRI

In the formation of stable $\alpha_2\beta_2$ hemoglobins, it is important to keep the concentrations of globin chains and heme balanced. Globin chains misfold and precipitate in the absence of proper binding to heme, as observed *in vitro* (93, 94) and *in vivo* with unstable hemoglobins caused by mutations that decrease heme incorporation (95, 96). Excess heme causes oxidative stress and is also cytotoxic. Biochemical and *in vitro* studies described above suggest that HRI may serve as a feedback inhibitor of globin synthesis to balance heme and globin synthesis by sensing heme availability as illustrated in Fig. 5. This hypothesis has been proven by targeted disruption of the HRI gene in mice (32).

To reveal the physiological function of HRI, heme deficiency was induced by feeding mice with an iron-deficient diet (32). The normal adaptive response to iron deficiency in humans and mice is the well-characterized microcytic hypochromic anemia with decreased mean corpuscular volume (MCV) and mean cell hemoglobin (MCH) in RBCs (Fig. 6). In $Hri^{-/-}$ mice, this physiological response to iron deficiency was dramatically altered. These mice develop a very unusual pattern of slight hyperchromic normocytic anemia with an accentuated decrease in RBC counts. Thus, HRI is critical in determining the size, number and hemoglobin content of the RBC. HRI is solely responsible for the adaptation of microcytic hypochromic anemia in iron deficiency.

Multiple variably sized globin inclusions were observed within reticulocytes and to a lesser extent within fully mature RBCs in iron-deficient $Hri^{-/-}$ mice as illustrated in Fig. 6. Together, studies using HRI knockout mice in iron deficiency uncovered the function of HRI in coordinating the synthesis of globins in RBC precursors with the concentration of heme *in vivo* as illustrated in Figs. 5 and 6. HRI normally ensures that no globin chains are translated in excess of what can be assembled into hemoglobin tetramers for the amount of heme available.

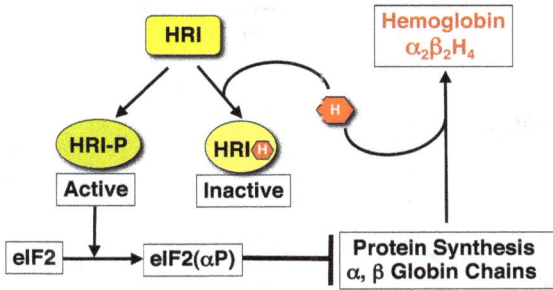

Fig. 5. HRI as a heme sensor to balance globin translation with heme concentrations. During the synthesis of hemoglobin, one molecule of heme is incorporated into each globin chain. When heme concentration is high, heme binds to HRI and keeps HRI in an inactive state, thereby permitting globin protein synthesis and the formation of stable hemoglobin. In heme deficiency, HRI is activated by autophosphorylation. Activated HRI phosphorylates eIF2α and inhibits globin protein synthesis by the mechanism illustrated in Fig. 1. HRI, therefore, acts as a heme sensor and a feedback inhibitor of globin synthesis to ensure that no globin is translated in excess of the heme available for assembly of stable hemoglobin. This figure is modified from the figure published originally in EMBO Journal. Han AP, Yu C, Lu L, Fujiwara Y, Browne C, Chin G, Fleming M, Leboulch P, Orkin SH, Chen JJ. 2001. Heme-regulated eIF2α kinase (HRI) is required for translational regulation and survival of erythroid precursors in iron deficiency. *EMBO J.* **20**: 6909–6918.

In the absence of HRI, heme-free globins precipitate in iron deficiency and become a major cell destruction component in the pathophysiology of iron deficiency anemia (32). Moreover, the survival rate of iron-deficient *Hri*$^{-/-}$ mice upon phenylhydrazine-induced hemolytic erythroid stress was dramatically reduced (32). *Hri*$^{-/-}$ RBCs were much more sensitive to additional stress such as oxidative stress induced by the administration of phenylhydrazine. This finding indicates that HRI plays a protective role in maintaining the integrity of mature red cells, particularly during combined assaults of iron deficiency together with additional stress.

The importance of HRI in the pathophysiology of heme deficiency disorders was further demonstrated by generating mice with combined deficiencies of HRI and ferrochelatase (*Fech*). *Fech-deficient* mice have a point mutation in Fech gene resulting in substitution of the amino acid at position 98 from Met to Lys. This mutation reduces the Fech enzyme activity

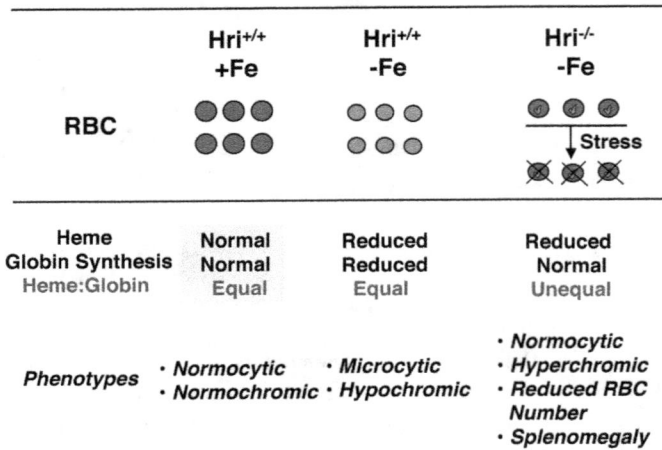

	Hri⁺/⁺ +Fe	Hri⁺/⁺ -Fe	Hri⁻/⁻ -Fe

Fig. 6. Necessity of HRI for the adaptation of microcytic hypochromic anemia in iron deficiency. In iron deficiency, heme concentration declines which leads to HRI activation and inhibition of globin synthesis as illustrated in Fig. 3. This results in decreased hemoglobin and total protein content in Wt RBCs. In the absence of HRI (*Hri⁻/⁻*), protein synthesis continues in the face of heme deficiency, resulting in excess globins. These heme-free globins are unstable and precipitate as inclusions (highlighted in green) in RBCs and their precursors, causing destruction of these cells. *Hri⁻/⁻* RBCs are of normal cell size, not microcytic, and slightly hyperchromic. However, there is significant decrease in the RBC number, reticulocytosis and splenomegaly in *Hri⁻/⁻* mice. This figure is modified from the figure published originally in EMBO Journal. Han AP, Yu C, Lu L, Fujiwara Y, Browne C, Chin G, Fleming M, Lebdouch P, Orkin SH, Chen JJ. 2001. Heme-regulated eIF2α kinase (HRI) is required for translational regulation and survival of erythroid precursors in iron deficiency. *EMBO J.* **20**: 6909–6918.

drastically to only 3%–6% of the Wt enzyme (97). Since Fech (the last enzyme of heme biosynthesis) inserts iron into protoporphyrin IX (PPIX) to form heme, *Fech-deficient* mice are functionally heme-deficient, and accumulate PPIX, similar to the human disease of erythropoietic proto-porphyria (EPP) (97, 98). HRI was activated in *Fech-deficient* reticulo-cytes, providing the *in vivo* evidence that HRI is regulated directly by heme and not by iron (56).

In the presence of HRI, *Fech-deficient* mice developed a microcytic hypochromic anemia. These mice became significantly more anemic in HRI deficiency. However, the MCV and MCH were not significantly decreased in HRI deficiency; the decreased hemoglobin was due to an overall decrease in the number of normochromic, normocytic RBCs, similar to the unusual effect of HRI deficiency on the normally microcytic anemia of iron deficiency (32). Furthermore, inclusion bodies were observed in reticulocytes from HRI deficient *Fech-deficient* animals. The morphologically and biochemically similar red cell abnormalities elicited by the absence of HRI both in iron and heme deficiencies further underscore the importance of HRI in inhibiting protein synthesis to avoid accumulation of excess heme-free globins in heme deficiency states.

Regulation of erythroid differentiation by HRI

Erythropoiesis in humans and mice is highly hierarchical (reviewed in (99)). The earliest committed erythroid progenitor identified is the burst-forming unit erythroid cell (BFU-E), followed by colony-forming unit cell (CFU-E). In the presence of erythropoietin, these erythroid progenitors grow and differentiate to proerythroblasts, basophilic, polychromic and orthochromic erythroblasts as illustrated in Fig. 7. The nuclei are then extruded upon maturation to reticulocytes. Finally erythropoiesis is completed with the elimination of mitochondria, ribosomes and other organelles to form mature erythrocytes.

Different stages of erythroid differentiation in mice can be monitored by expression of CD71 transferrin receptor and erythroid-specific cell surface Ter119 protein using fluorescent activated cell sorting (FACS) analysis (100, 101). As erythroid progenitors differentiate, there is a transient increase of CD71 expression and increase of Ter119 expression at the basophilic erythroblast stage. Thus, BFU-E and CFU-E cells are CD71^{low}Ter119$^-$, proerythroblasts are CD71^{high}Ter119$^-$; basophilic erythroblasts are CD71^{high}Ter119$^+$; polychromic and orthochromic erythroblasts are CD71^{medium}Ter119$^+$ (Fig. 7).

In addition to accumulation of globin inclusions in reticulocytes and RBCs, $Hri^{-/-}$ mice also develop ineffective erythropoiesis during iron deficiency with marked erythroid expansion and increased apoptosis of erythroid precursors in the bone marrow and spleen (32). Liu *et al.* have reported recently that erythroid differentiation of E14.5 fetal cells in iron deficiency was inhibited at the basophilic erythroblast stage, and this inhibition was more severe in HRI deficiency (102). Furthermore, microarray gene profiling analysis demonstrates that HRI is required for adaptive gene expression in erythroid precursors during chronic iron deficiency. The

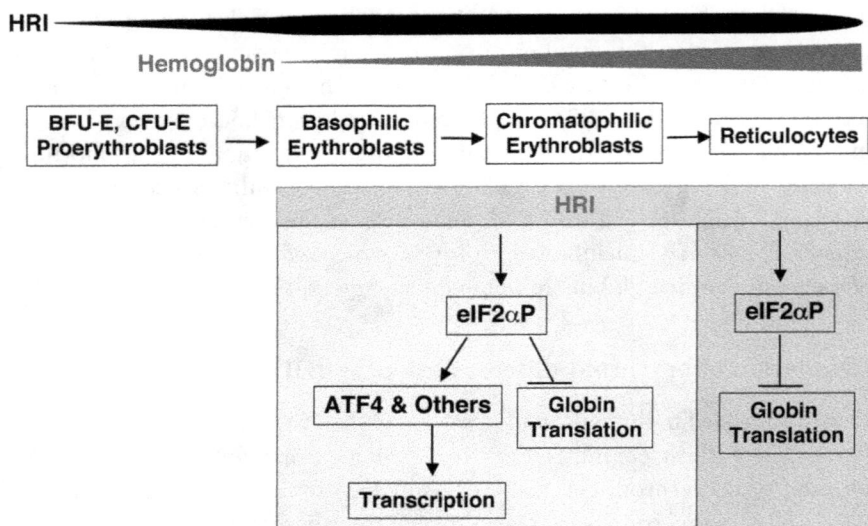

Fig. 7. Dual roles of HRI in erythropoiesis. During erythroid differentiation, HRI expression increases starting with basophilic erythroblasts through reticulocytes. In reticulocytes, the role of HRI is limited to translational control only because there are no nuclei in these cells. In the nucleated precursors, the role of HRI extends beyond translational control to transcriptional control by the selective translation of transcriptional factors such as ATF4.

number of genes with expression affected more than 2-fold increased from 213 in iron deficiency and 73 in HRI deficiency to 3135 in combined iron and HRI deficiencies. Many of these genes are regulated by master erythroid transcriptional factors, Gata1 and Fog1 (102). These results demonstrate the necessity of HRI in adaptation to erythropoiesis during iron deficiency.

Examination of HRI expression in sorted populations of fetal liver cells at different stages of erythroid differentiation shows that HRI expression, both at mRNA and protein levels, increases during erythroid differentiation with higher expression in Ter119[+] erythroblasts (102). This higher expression of HRI in erythroblasts, in which globin proteins are actively made in large quantities, is consistent with the known function of HRI in regulating globin translation (Fig. 7). The contribution of HRI to eIF2α phosphorylation among the family of eIF2α kinases is also increased dramatically during erythroid differentiation from 35% in

Ter119⁻ cells to over 90% in CD71^{med}Ter119⁺ cells (102). This increased expression of HRI starting in basophilic erythroblasts also provides supporting evidence for the role of HRI in erythroid differentiation, especially under the stress condition of iron deficiency.

Integrated stress response of eIF2αP signaling

In addition to inhibition of global protein synthesis, the second important consequence of eIF2α phosphorylation is the reprogramming of the translation and transcription of genes required for stress response, which is termed the "Integrated Stress Response", ISR ((103), and reviewed in (104, 105)). Under ISR, translation of some transcriptional factors is selectively increased amid the inhibition of general protein synthesis (Fig. 8). The best example of this type of translational up-regulation has been

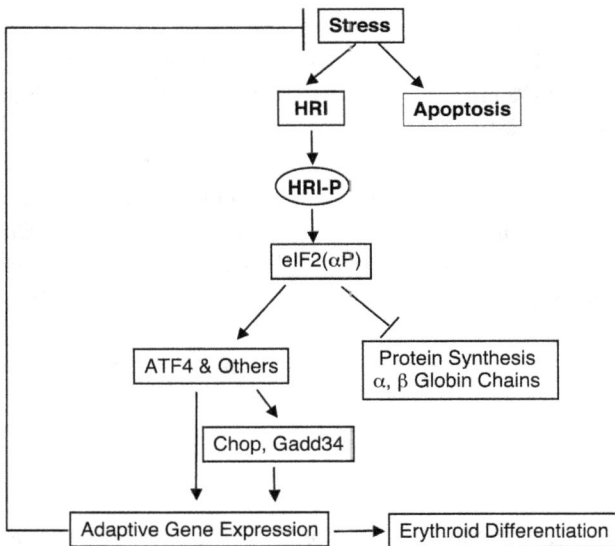

Fig. 8. The HRI signaling pathway in stress response and erythroid differentiation. Upon stress in erythroid cells, HRI is activated first to inhibit globin synthesis to reduce toxic globin precipitates. Importantly, HRI also participates in the second arm of defense by modulating gene expression necessary for adaptation to stress conditions or apoptosis if the stress is insurmountable. Additionally, this HRI signaling pathway may also be required for erythroid differentiation especially under stress conditions.

documented in amino acid starvation of yeast in which GCN2 is activated and the expression of the transcriptional factor GCN4 is up-regulated to induce the enzymes of amino acid biosynthetic pathways. This up-regulation requires the presence of upstream open reading frames (uORFs) in the 5-UTR of the GCN4 mRNA (reviewed in (21)).

In mammalian cells, transcriptional factor ATF4 (activating transcriptional factor 4, the metazoan homologue of GCN4) is up-regulated by PERK upon ER stress and by GCN2 upon amino acid starvation (41, 106). Like GCN4, ATF4 also contains uORFs, which are essential for its translational activation by eIF2α phosphorylation (107, 108). As illustrated in Fig. 8, a major target gene of ATF4 is the transcriptional factor, C/EBP homologous protein-10 (Chop, also known as growth arrest and differentiation protein 153, GADD153), which is up-regulated in a wide variety of cells upon many stresses (109, 110). In ER stress, induction of Chop leads to expression of GADD34, which recruits eIF2αP for dephosphorylation by PPase 1 (24–27). This action of GADD34 in regenerating active eIF2 (Fig. 2) is necessary for the recovery of protein synthesis of stress-induced gene expression that occurs late in stress response (28). It is important to note that the ATF4 pathway only accounts for about 50% of gene expression in PERK-mediated ER stress response (103). This finding indicates that there are additional target mRNAs that are up-regulated translationally by eIF2αP during ER stress. Thus, while many components have been recently identified in ER unfolded protein response, the PERK/eIF2αP signaling pathway is still not well understood. Recently, translations of C/EBP α and β have also been shown to be regulated by eIF2α phosphorylation (111).

While HRI plays an important role in balancing heme and globin synthesis in erythroid cells, its function is not limited to heme deficiency. HRI is also activated by sodium arsenite-induced oxidative stress, osmotic shock and heat shock, but not by ER stress, amino acid or serum starvation (68). Furthermore, HRI is the principal eIF2α kinase responsible for protection against oxidative stress, heat shock and UV irradiation of erythroid cells. In addition to eIF2α phosphorylation and inhibition of protein synthesis (68), arsenite treatment also induced expression of ATF4, Chop and Gadd34 in $Hri^{+/+}$ erythroid precursor cells, but not in $Hri^{-/-}$ cells (112). Thus, ISR is also operative in erythroid precursors upon oxidative stress induced by arsenite and importantly requires HRI. These results indicate that the HRI–ISR signaling pathway may be necessary for inducing the transcription of genes required for stress erythropoiesis starting at the basophilic erythroblast stage as illustrated in Fig. 7.

Interestingly, components of the HRI stress response have been shown to be involved in erythropoiesis. Knockout of the ATF4 gene in mice results in growth retardation and transient fetal anemia with relatively normal erythropoiesis in adulthood (113). Chop expression was up-regulated in mouse erythroleukemic cells upon induction of erythroid differentiation by erythropoietin or DMSO (114). Furthermore, GADD34 is required for hemoglobin synthesis (115). $Gadd34^{-/-}$ mice developed a mild microcytic hypochromic anemia and had a higher level of eIF2αP in their reticulocytes, establishing the function of GADD34 in the dephosphorylation of eIF2αP in erythroid cells.

A recent study of the targeted disruption of the CReP gene, the constitutively expressed regulator of eIF2αP dephosphorylation (116), also revealed the role of eIF2αP dephosphorylation in erythropoiesis and development. $CReP^{-/-}$ mice had severe growth retardation, impaired erythropoiesis and postnatal death on the first day (29). It is to be noted that the anemia of $CReP^{-/-}$ embryos cannot be rescued by deletion of HRI or PERK alone (29), suggesting that more than one eIF2α kinase is necessary for the proliferation of erythroid progenitors in the fetal liver. Furthermore, both inadequate eIF2αP signaling (as in the case of ATF4 deficiency) and excessive eIF2αP signaling (as in the case of CReP deficiency) are associated with fetal anemia. These findings underscore the importance of the fine regulation of the level of eIF2αP and the subsequent signaling events for the proliferation and differentiation of fetal liver erythroid progenitors.

Concluding remarks

Heme not only serves as a prosthetic group of hemoglobin, the most abundant protein in RBCs, but also plays critical roles in the biosynthesis of globins and erythropoiesis. Translational control by HRI is one major mechanism employed by heme to regulate these processes. HRI is a heme sensor protein with two distinct heme-binding domains to gauge intracellular heme concentrations. HRI is necessary for heme-regulated translation in erythroid precursors. In heme deficiency, HRI is activated and translation is inhibited by eIF2αP. In this manner, HRI serves as a feedback inhibitor for globin translation to coordinate globin synthesis with heme availability. Additionally, HRI also activates the eIF2αP ISR signaling pathway for adaptive gene expression to mitigate stress. The role of the eIF2αP signaling pathway in the regulation of erythropoiesis by HRI and other eIF2α kinases is currently an active research area. In conclusion, HRI plays dual roles during erythropoiesis: coordination of heme and globin synthesis, and activation of gene expression via induction of ATF4 and other yet-to-be-identified erythroid transcriptional factors for survival and differentiation of erythroid precursors especially under stress conditions. Beyond the critical role in heme deficiency, HRI is also essential to reduce the phenotypic severity of β-thalassemia. Lacking both copies of HRI is embryonically lethal while lacking one copy also resulted in more severe symptoms in mice (56). Thus, HRI and its signaling pathways may be potential novel pharmaceutical targets to combat thalassemia and other red cell disorders.

Acknowledgement

The research performed in my laboratory was supported in part by grants from National Institutes of Health, National Institute of Diabetes and Digestive & Kidney Diseases (DK16272 and DK 88442). Dr. Suragani was supported by a postdoctoral fellowship from the Cooley's Anemia Foundation from July 2007 to June 2009.

References

1. Andrews NC. 2000. Iron metabolism: Iron deficiency and iron overload. *Annu Rev Genomics Hum Genet* 1: 75–98.
2. Iolascon A, De Falco L, Beaumont C. 2009. Molecular basis of inherited microcytic anemia due to defects in iron acquisition or heme synthesis. *Haematologica* 94: 395–408.
3. Kruh J, Borsook G. 1956. Hemoglobin synthesis in rabbit reticulocytes *in vitro*. *J Biol Chem* 220: 905–915.
4. Kassenaar A, Morell H, London IM. 1957. The incorporation of glycine into globin and the synthesis of heme *in vitro* in duck erythrocytes. *J Biol Chem* 229: 423–435.
5. Morell H, Savoie JC, London IM. 1958. The biosynthesis of heme and the incorporation of glycine into globin in rabbit bone marrow *in vitro*. *J Biol Chem* 233: 923–929.
6. Bruns GP, London IM. 1965. The effect of hemin on the synthesis of globin. *Biochem Biophys Res Comm* 18: 236–242.

7. Grayzel AIP, Horchner P, London IM. 1966. The stimulation of globin synthesis by heme. *Proc Natl Acad Sci U S A* 55: 650–655.

8. Igarashi K, Sun J. 2006. The heme-Bach1 pathway in the regulation of oxidative stress response and erythroid differentiation. *Antioxid Redox Signal* 8: 107–118.

9. Waxman HS, Rabinovitz M. 1965. Iron supplementation *in vitro* and the state of aggregation and function of reticulocyte ribosomes in hemoglobin synthesis. *Biochem Biophys Res Commun* 19: 538–545.

10. Waxman HS, Rabinovitz M. 1966. Control of reticulocyte polyribosome content and hemoblogin synthesis by heme. *Biochim Biophys Acta* 129: 369–379.

11. Zucker WV, Schulman HM. 1968. Stimulation of globin-chain initiation by hemin in the reticulocyte cell-free system. *Proc Natl Acad Sci U S A* 59: 582–589.

12. Legon S, Jackson RJ, Hunt T. 1973. Control of protein synthesis in reticulocyte lysates by haemin. *Nat New Biol* 241: 150–152.

13. Ernst V, Levin DH, London IM. 1979. *In situ* phosphorylation of the α subunit of eukaryotic initiation factor 2 in reticulocyte lysates inhibited by heme deficiency, double-stranded RNA, oxidized glutathione, or the heme-regulated protein kinase. *Proc Natl Acad Sci U S A* 76: 2118–2122.

14. Leroux A, London IM. 1982. Regulation of protein synthesis by phosphorylation of eukaryotic initiation factor 2α in intact reticulocytes and reticulocyte lysates. *Proc Natl Acad Sci U S A* 79: 2147–2151.

15. Maxwell CR, Rabinovitz M. 1969. Evidence for an inhibitor in the control of globin synthesis by hemin in a reticulocyte lysate. *Biochem Biophys Res Commun* 35: 79–85.

16. Adamson SD, Herbert E, Godchaux W. 1968. Factors affecting the rate of protein synthesis in lysate systems from reticulocytes. *Arch Biochem Biophys* 125: 671–683.

17. Howard GA, Adamson SD, Herbert E. 1970. Studies on cessation of protein synthesis in reticulocyte lysate cell-free system. *Biochim Biophys Acta* 213: 237–240.

18. Levin DH, Ranu RS, Ernst V, London IM. 1976. Regulation of protein synthesis in reticulocyte lysates: Phosphorylation of methionyl-tRNA$_f$ binding factor by protein kinase activity of the translational inhibitor isolated from heme-deficient lysates. *Proc Natl Acad Sci U S A* 73: 3112–3116.

19. Kramer G, Cimadevilla M, Hardesty B. 1976. Specificity of the protein kinase activity associated with the hemin-controlled repressor of rabbit reticulocyte. *Proc Natl Acad Sci U S A* 73: 3078–3082.

20. Chen J-J. 2000. Heme-regulated eIF-2α kinase. In *Translational Control of Gene Expression*, eds. N Sonenberg, JWB Hershey, MB Mathews, pp. 529–546. Cold Springs Harbor: Cold Spring Harbor Laboratory Press.

21. Hinnebusch AG, Dever, TE, Asano K. 2007. Mechanism of translational initiation in the yeast *Saccharomyces cerevisiae*. In *Translational Control in Biology and Medicine*, eds. MB Mathews, N Sonenberg, JWB Hershey, pp. 225–268. Cold Spring Harbor: Cold Spring Harbor Laboratory Press.

22. Hershey JW, Merrick WC. 2000. Pathway and mechanism of initiation of protein synthesis. In: *Translational Control*, eds. N Sonenberg, JW Hershey, M Matthews, pp. 33–88. Cold Spring Harbor: Cold Spring Harbor Laboratory Press.

23. Krishnamoorthy T, Pavitt GD, Zhang F, Dever TE, Hinnebusch AG. 2001. Tight binding of the phosphorylated alpha subunit of initiation factor 2 (eIF2alpha) to the regulatory subunits of guanine nucleotide exchange factor eIF2B is required for inhibition of translation initiation. *Mol Cell Biol* 21: 5018–5030.

24. Novoa I, Zeng H, Harding HP, Ron D. 2001. Feedback inhibition of the unfolded protein response by GADD34-mediated dephosphorylation of eIF2alpha. *J Cell Biol* 153: 1011–1022.

25. Kojima E, Takeuchi A, Haneda M, Yagi A, Hasegawa T, Yamaki K, Takeda K, Akira S, Shimokata K, Isobe K. 2003. The function of GADD34 is a recovery from a shutoff of protein synthesis induced by ER stress: Elucidation by GADD34-deficient mice. *Faseb J* 17: 1573–1575.

26. Connor JH, Weiser DC, Li S, Hallenbeck JM, Shenolikar S. 2001. Growth arrest and DNA damage-inducible protein GADD34 assembles a novel signaling complex containing protein phosphatase 1 and inhibitor 1. *Mol Cell Biol* 21: 6841–6850.

27. Brush MH, Weiser DC, Shenolikar S. 2003. Growth arrest and DNA damage-inducible protein GADD34 targets protein phosphatase 1 alpha to the endoplasmic reticulum and promotes dephosphorylation of the alpha subunit of eukaryotic translation initiation factor 2. *Mol Cell Biol* 23: 1292–1303.

28. Novoa I, Zhang Y, Zeng H, Jungreis R, Harding HP, Ron D. 2003. Stress-induced gene expression requires programmed recovery from translational repression. *EMBO J* 22: 1180–1187.

29. Harding HP, Zhang Y, Scheuner D, Chen JJ, Kaufman RJ, Ron D. 2009. Ppp1r15 gene knockout reveals an essential role for translation initiation factor 2 alpha (eIF2alpha) dephosphorylation in mammalian development. *Proc Natl Acad Sci U S A* 106: 1832–1837.

30. Gross M, Redman R. 1987. Effect of antibody to the hemin-controlled translational repressor in rabbit reticulocyte. *Biochim Biophys Acta* 908: 123–130.
31. Pal JK, Chen J-J, London IM. 1991. Tissue distribution and immunoreactivity of heme-regulated eIF-2α kinase determined by monoclonal antibodies. *Biochem* 30: 2555–2562.
32. Han AP, Yu C, Lu L, Fujiwara Y, Browne C, Chin G, Fleming M, Leboulch P, Orkin SH, Chen JJ. 2001. Heme-regulated eIF2alpha kinase (HRI) is required for translational regulation and survival of erythroid precursors in iron deficiency. *EMBO J* 20: 6909–6918.
33. Ron D, Harding HP. 2007. eIF2α phosphorylation in cellular stress responses and disease. In: *Translational Control in Biology and Medicine*, eds. MB Mathews, N Sonenberg, JWB Hershey, pp. 345–368. Cold Spring Harbor: Cold Spring Harbor Laboratory Press.
34. Lu PD, Jousse C, Marciniak SJ, Zhang Y, Novoa I, Scheuner D, Kaufman RJ, Ron D, Harding HP. 2004. Cytoprotection by pre-emptive conditional phosphorylation of translation initiation factor 2. *EMBO J* 23: 169–179.
35. Chen J-J, Throop MS, Gehrke L, Kuc I, Pal JK, Brodsky M, London IM. 1991. Cloning of the cDNA of the heme-regulated eukaryotic initiation factor 2α (eIF-2α) kinase of rabbit reticulocytes: Homology to yeast GCN2 protein kinase and human double-stranded-RNA-dependent eIF-2α kinase. *Proc Natl Acad Sci U S A* 88: 7729–7733.
36. Meurs E, Chong K, Galabru J, Thomas NSB, Kerr IM, Williams BRG, Hovanessian AG. 1990. Molecular cloning and characterization of human double-stranded RNA activated protein kinase induced by interferon. *Cell* 62: 379–390.
37. Ramirez M, Wek RC, Hinnebusch AG. 1991. Ribosome association of GCN2 protein kinase, a translational activation of the GCN4 gene of *Saccharomyces cerevisiae*. *Mol Cell Biol* 11: 3027–3036.
38. Chong KL, Schappert K, Meurs E, Feng F, Donahue TF, Friesen JD, Hovanessian AG, Williams BRG. 1992. Human p68 kinase exhibits growth suppression in yeast and homology to the translational regulator GCN2. *EMBO J.* 11: 1553–1562.
39. Berlanga JJ, Herrero S, de Haro C. 1998. Characterization of the hemin-sensitive eukaryotic initiation factor 2α kinase from mouse nonerythroid cells. *J Biol Chem* 273: 32340–32346.
40. Shi Y, Vattem KM, Sood R, An J, Liang J, Stramm L, Wek RC. 1998. Identification and characterization of pancreatic eukaryotic initiation factor 2 α-subunit kinase, PEK, involved in translational control. *Mol Cell Biol* 18: 7499–7509.

41. Harding HP, Zhang Y, Ron D. 1999. Protein translation and folding are coupled by an endoplasmic reticulum-resident kinase. *Nature* 397: 271–274.

42. Colthurst DR, Campbell DG, Proud CG. 1987. Structure and regulation of eukaryotic initiation factor eIF2: Sequence of the site in the alpha subunit phosphorylated by the haem-controlled repressor and by the double-stranded RNA-activated inhibitor. *Eur J Biochem* 166: 357–363.

43. Dever TE, Chen J-J, Barber GN, Cigan AM, Feng L, Donahue TF, London IM, Katze MG, Hinnebusch AG. 1993. Mammalian eIF-2α kinases functionally substitute for GCN2 in the GCN4 translational control mechanism of yeast. *Proc Natl Acad Sci U S A* 90: 4616–4620.

44. Kaufman RJ. 2000. Double-stranded RNA-activated protein kinase PKR. In *Translational Control of Gene Expression*, eds. N Sonenberg, JWB Hershey, MB Matthews, pp. 503–528. Cold Spring Harbor: Cold Spring Harbor Laboratory Press.

45. Hinnebusch AG. 1996. Translational control of GCN4: gene-specific regulation by phosphorylation of eIF2. In *Translational Control*, eds. JWB Hershey, MB Mathews, N Sonenberg, pp. 199–244. Cold Spring Harbor: Cold Spring Harbor Laboratory Press.

46. Ron D, Harding HP. 2000. PERK and translational control by stress in endoplasmic reticulum. In *Translational Control of Gene Expression*, eds. N Sonenberg, JWB Hershey, MB Mathews, pp. 547–560. Cold Springs Harbor: Cold Spring Harbor Laboratory Press.

47. Crosby JS, Lee K, London IM, Chen J-J. 1994. Erythroid expression of the heme-regulated eIF-2α kinase. *Mol Cell Biol* 14: 3906–3914.

48. Liu S, Bhattacharya S, Han A, Suragani RN, Zhao W, Fry RC, Chen JJ. 2008. Haem-regulated eIF2alpha kinase is necessary for adaptive gene expression in erythroid precursors under the stress of iron deficiency. *Br J Haematol* 143: 129–137.

49. Yang Y-L, Reis LFL, Pavlovic J, Aguzzi A, Schafer R, Kumar A, Williams BRG, Aguet M, Weissman C. 1995. Deficient signaling in mice devoid of double-stranded RNA-dependent protein kinase. *EMBO J* 14: 6095–6106.

50. Abraham N, Stojdl DF, Duncan PI, Methot N, Ishii T, Dube M, Vanderhyden BC, Atkins HL, Gray DA, McBurney MW, Koromilas AE, Brown EG, Sonenberg N, Bell JC. 1999. Characterization of transgenic mice with targeted disruption of the catalytic domain of the double-stranded RNA-dependent protein kinase, PKR. *J Biol Chem* 274: 5953–5962.

51. Harding HP, Zeng H, Zhang Y, Jungries R, Chung P, Plesken H, Sabatini DD, Ron D. 2001. Diabetes mellitus and exocrine pancreatic dysfunction in perk−/− mice reveals a role for translational control in secretory cell survival. *Mol Cell* 7: 1153–1163.

52. Zhang P, McGrath BC, Reinert J, Olsen DS, Lei L, Gill S, Wek SA, Vattem KM, Wek RC, Kimball SR, Jefferson LS, Cavener DR. 2002. The GCN2 eIF2alpha kinase is required for adaptation to amino acid deprivation in mice. *Mol Cell Biol* 22: 6681–6688.

53. Hao S, Sharp JW, Ross-Inta CM, McDaniel BJ, Anthony TG, Wek RC, Cavener DR, McGrath BC, Rudell JB, Koehnle TJ, Gietzen DW. 2005. Uncharged tRNA and sensing of amino acid deficiency in mammalian piriform cortex. *Science* 307: 1776–1778.

54. Maurin AC, Jousse C, Averous J, Parry L, Bruhat A, Cherasse Y, Zeng H, Zhang Y, Harding HP, Ron D, Fafournoux P. 2005. The GCN2 kinase biases feeding behavior to maintain amino acid homeostasis in omnivores. *Cell Metab* 1: 273–277.

55. Costa-Mattioli M, Gobert D, Stern E, Gamache K, Colina R, Cuello C, Sossin W, Kaufman R, Pelletier J, Rosenblum K, Krnjevic K, Lacaille JC, Nader K, Sonenberg N. 2007. eIF2alpha phosphorylation bidirectionally regulates the switch from short- to long-term synaptic plasticity and memory. *Cell* 129: 195–206.

56. Han AP, Fleming MD, Chen JJ. 2005. Heme-regulated eIF2alpha kinase modifies the phenotypic severity of murine models of erythropoietic protoporphyria and beta-thalassemia. *J Clin Invest* 115: 1562–1570.

57. Liu S, Suragani RN, Wang F, Han A, Zhao W, Andrews NC, Chen JJ. 2007. The function of heme-regulated eIF2alpha kinase in murine iron homeostasis and macrophage maturation. *J Clin Invest* 117: 3296–3305.

58. Scheuner D, Song B, McEwen E, Liu C, Laybutt R, Gillespie P, Saunders T, Bonner-Weir S, Kaufman RJ. 2001. Translational control is required for the unfolded protein response and *in vivo* glucose homeostasis. *Mol Cell* 7: 1165–1176.

59. Gross M, Rabinovitz M. 1972. Control of globin synthesis by heme: Factors influencing formation of an inhibitor of globin chain initiation in reticulocyte lysates. *Biochim Biophys Acta* 287: 340–352.

60. Hunt T. 1979. The control of protein synthesis in rabbit reticulocyte lysates. In *Miami Winter Symposium: From Gene to Protein*, eds. TR Russel, K Brew, J Schultz, H Haber, pp. 321–345. New York: Academic Press.

61. Yang JM, London IM, Chen J-J. 1992. Effects of hemin and porphyrin compounds on intersubunit disulfide formation of heme-regulated eIF-2α kinase and the regulation of protein synthesis in reticulocyte lysates. *J Biol Chem* 267: 20519–20524.

62. Gross M, Mendelewski J. 1978. An association between the formation of the hemin-controlled translational repressor and the phosphorylation of a 100,000 molecular weight protein. *Biochim Biophys Acta* 520: 650–663.

63. Fagard R, London IM. 1981. Relationship between the phosphorylation and activity of the heme-regulated eIF2α kinase. *Proc Natl Acad Sci U S A* 78: 866–870.

64. Chefalo P, Oh J, Rafie-Kolpin M, Kan B, Chen J-J. 1998. Heme-regulated eIF2α kinase purifies as a hemoprotein. *Eur J Biochem* 258: 820–830.

65. Bauer BN, Rafie-Kolpin M, Lu L, Han A, Chen J-J. 2001. Multiple autophosphorylation is essential for the formation of the active and stable homodimer of heme-regulated eIF2α kinase. *Biochemistry* 40: 11543–11551.

66. Rafie-Kolpin M, Han AP, Chen JJ. 2003. Autophosphorylation of threonine 485 in the activation loop is essential for attaining eIF2alpha kinase activity of HRI. *Biochemistry* 42: 6536–6544.

67. Chefalo PJ, Yang JM, Ramaiah KVA, Gehrke L, Chen JJ. 1994. Inhibition of protein synthesis in insect cells by baculovirus-expressed heme-regulated eIF-2α kinase. *J Biol Chem* 269: 25788–25794.

68. Lu L, Han AP, Chen JJ. 2001. Translation initiation control by heme-regulated eukaryotic initiation factor 2alpha kinase in erythroid cells under cytoplasmic stresses. *Mol Cell Biol* 21: 7971–7980.

69. Trachsel H, Ranu RS, London IM. 1978. Regulation of protein synthesis in rabbit reticulocyte lysates: Purification and characterization of heme-reversible translational inhibitor. *Proc Natl Acad Sci U S A* 75: 3654–3658.

70. Chen J-J, Pal JK, Petryshyn R, Kuo I, Yang JM, Throop MS, Gehrke L, London IM. 1991. Amino acid microsequencing of the internal tryptic peptides of heme-regulated eukaryotic initiation factor 2a subunit kinase: Homology to protein kinases. *Proc Natl Acad Sci U S A* 88: 315–319.

71. Chen J-J, Yang JM, Petryshyn R, Kosower N, London IM. 1989. Disulfide bond formation in the regulation of eIF-2α kinase by heme. *J Biol Chem* 264: 9559–9564.

72. Asseffa A, Smith SJ, Nagata K, Gillette J, Gelboin HV, Gonzalez FJ. 1989. Novel exogenous heme-dependent expression of mammalian cytochrome P450 using baculovirus. *Arch Biochem Biophys* 274: 481–490.

73. Richards MK, Marletta MA. 1994. Characterization of neuronal nitric oxide synthase and a C415H mutant, purified from a baculovirus overexpression system. *Biochemistry* 33: 14723–14732.

74. Rafie-Kolpin M, Chefalo PJ, Hussain Z, Hahn J, Uma S, Matts RL, Chen J-J. 2000. Two heme-binding domains of heme-regulated eIF-2α kinase: N-terminus and kinase insertion. *J Biol Chem* 275: 5171–5178.

75. Inuzuka T, Yun BG, Ishikawa H, Takahashi S, Hori H, Matts RL, Ishimori K, Morishima I. 2004. Identification of crucial histidines for heme binding in

the N-terminal domain of the heme-regulated eIF2alpha kinase. *J Biol Chem* 279: 6778–6782.

76. Igarashi J, Sato A, Kitagawa T, Ycshimura T, Yamauchi S, Sagami I, Shimizu T. 2004. Activation of heme-regulated eukaryotic initiation factor 2alpha kinase by nitric oxide is induced by the formation of a five-coordinate NO-heme complex: Optical absorption, electron spin resonance, and resonance raman spectral studies. *J Biol Chem* 279: 15752–15762.

77. Yun BG, Matts JA, Matts RL. 2005. Interdomain interactions regulate the activation of the heme-regulated eIF 2 alpha kinase. *Biochim Biophys Acta* 1725: 174–181.

78. Igarashi J, Murase M, Iizuka A, Pichierri F, Martinkova M, Shimizu T. 2008. Elucidation of the heme binding site of heme-regulated eukaryotic initiation factor 2alpha kinase and the role of the regulatory motif in heme sensing by spectroscopic and catalytic studies of mutant proteins. *J Biol Chem* 283: 18782–18791.

79. Lathrop JT, Timko MP. 1993. Regulation by heme of mitochondrial protein transport through a conserved amino acid motif. *Science* 259: 522–525.

80. Zhang L, Guarente L. 1995. Heme binds to a short sequence that serves a regulatory function in diverse proteins. *EMBO J* 14: 313–320.

81. Chen J-J, London IM. 1995. Regulation of protein synthesis by heme-regulated eIF-2α kinase. *TIBS* 20: 105–108.

82. Christiansen JH, Coles EG, Robinson V, Pasini A, Wilkinson DG. 2001. Screening from a subtracted embryonic chick hindbrain cDNA library: Identification of genes expressed during hindbrain, midbrain and cranial neural crest development. *Mech Dev* 102: 119–133.

83. Zhu R, Zhang YB, Chen YD, Dong CW, Zhang FT, Zhang QY, Gui JF. 2006. Molecular cloning and stress-induced expression of paralichthys olivaceus heme-regulated initiation factor 2alpha kinase. *Dev Comp Immunol* 30: 1047–1059.

84. Welch JJ, Watts JA, Vakoc CR, Yao Y, Wang H, Hardison RC, Blobel GA, Chodosh LA, Weiss MJ. 2004. Global regulation of erythroid gene expression by transcription factor GATA-1. *Blood* 104: 3136–3147.

85. Petryshyn R, Rosa F, Fagard R, Levin D, London IM. 1984. Control of protein synthesis in human reticulocytes by heme-regulated and double-stranded RNA dependent eIF-2 alpha kinases. *Biochem Biophys Res Commun* 119: 891–899.

86. Mellor H, Flowers KM, Kimball SR, Jefferson LS. 1994. Cloning and characterization of cDNA encoding rat hemin-sensitive initiation factor-2α (eIF-2α) kinase. *J Biol Chem* 269: 10201–10204

87. Uma S, Yun BG, Matts RL. 2001. The heme-regulated eukaryotic initiation factor 2alpha kinase. A potential regulatory target for control of protein synthesis by diffusible gases. *J Biol Chem* 276: 14875–14883.

88. McEwen E, Kedersha N, Song B, Scheuner D, Gilks N, Han A, Chen JJ, Anderson P, Kaufman RJ. 2005. Heme-regulated inhibitor kinase-mediated phosphorylation of eukaryotic translation initiation factor 2 inhibits translation, induces stress granule formation, and mediates survival upon arsenite exposure. *J Biol Chem* 280: 16925–16933.

89. Liao M, Pabarcus MK, Wang Y, Hefner C, Maltby DA, Medzihradszky KF, Salas-Castillo SP, Yan J, Maher JJ, Correia MA. 2007. Impaired dexamethasone-mediated induction of tryptophan 2,3-dioxygenase in heme-deficient rat hepatocytes: translational control by a hepatic eIF2alpha kinase, the heme-regulated inhibitor. *J Pharmacol Exp Ther* 323: 979–989.

90. Zhan K, Vattem KM, Bauer BN, Dever TE, Chen JJ, Wek RC. 2002. Phosphorylation of eukaryotic initiation factor 2 by heme-regulated inhibitor kinase-related protein kinases in *Schizosaccharomyces pombe* is important for fesistance to environmental stresses. *Mol Cell Biol* 22: 7134–7146.

91. Zhan K, Narasimhan J, Wek RC. 2004. Differential activation of eIF2 kinases in response to cellular stresses in *Schizosaccharomyces pombe*. *Genetics* 168: 1867–1875.

92. Mohrle JJ, Zhao Y, Wernli B, Franklin RM, Kappes B. 1997. Molecular cloning, characterization and localization of PfPK4, an eIF2α kinase-related enzyme from the malarial parasite *Plasmodium falciparum*. *Biochem J* 328: 677–687.

93. Waks M, Yip YK, Beychok, S. 1973. Influence of prosthetic groups on protein folding and subunit assembly: Recombination of separated human alpha-and beta-globin chains with heme and alloplex interactions of globin chains with heme-containing subunits. *J Biol Chem* 248: 6462–6470.

94. Yip YK, Waks M, Beychok, S. 1972. Influence of prosthetic groups on protein folding and subunit assembly. I. Conformational differences between separated human alpha- and beta-globins. *J Biol Chem* 247: 7237–7244.

95. Dacie JV, Shinton NK, Gaffney PJ, Lehmann H. 1967. Haemoglobin Hammersmith (beta-42 (CDI) Phe replaced by Ser). *Nature* 216: 663–665.

96. Wajcman H, Kister J, Vasseur C, Blouquit Y, Trastour JC, Cottenceau D, Galacteros F. 1992. Structure of the EF corner favors deamidation of asparaginyl residues in hemoglobin: The example of Hb La Roche-sur-Yon [beta 81 (EF5) Leu–His]. *Biochim Biophys Acta* 1138: 127–132.

97. Tutois S, Montagutelli X, Da Silva V, Jouault H, Rouyer-Fessard P, Leroy-Viard K, Guenet JL, Nordmann Y, Beuzard Y, Deybach JC. 1991. Erythropoietic protoporphyria in the house mouse. A recessive inherited

ferrochelatase deficiency with anemia, photosensitivity, and liver disease. *J Clin Invest* 88: 1730–1736.

98. Boulechfar S, Lamoril J, Montagutelli X, Guenet JL, Deybach JC, Nordmann Y, Dailey H, Grandchamp B, de Verneuil H. 1993. Ferrochelatase structural mutant (Fechm1Pas) in the house mouse. *Genomics* 16: 645–648.

99. Spivak JL. 2005. The anaemia of cancer: Death by a thousand cuts. *Nat Rev Cancer* 5: 543–555.

100. Zhang J, Socolovsky M, Gross AW, Lodish HF. 2003. Role of Ras signaling in erythroid differentiation of mouse fetal liver cells: Functional analysis by a flow cytometry-based novel culture system. *Blood* 102: 3938–3946.

101. Socolovsky M. 2007. Molecular insights into stress erythropoiesis. *Curr Opin Hematol* 14: 215–224.

102. Liu S, Bhattacharya S, Han A, Suragani RN, Zhao W, Fry RC, Chen JJ. 2008. Haem-regulated eIF2alpha kinase is necessary for adaptive gene expression in erythroid precursors under the stress of iron deficiency. *Br J Haematol* 143: 129–137.

103. Harding HP, Zhang Y, Zeng H, Novoa I, Lu PD, Calfon M, Sadri N, Yun C, Popko B, Paules R, Stojdl DF, Bell JC, Hettmann T, Leiden JM, Ron D. 2003. An integrated stress response regulates amino acid metabolism and resistance to oxidative stress. *Mol Cell* 11: 619–633.

104. Holcik M, Sonenberg N. 2005. Translational control in stress and apoptosis. *Nat Rev Mol Cell Biol* 6: 318–327.

105. Ron D, Walter P. 2007. Signal integration in the endoplasmic reticulum unfolded protein response. *Nat Rev Mol Cell Biol* 8: 519–529.

106. Harding HP, Novoa II, Zhang Y, Zeng H, Wek R, Schapira M, Ron D. 2000. Regulated translation initiation controls stress-induced gene expression in mammalian cells. *Mol Cell* 6: 1099–1103.

107. Lu PD, Harding HP, Ron D. 2004. Translation reinitiation at alternative open reading frames regulates gene expression in an integrated stress response. *J Cell Biol* 167: 27–33.

108. Vattem KM, Wek RC. 2004. Reinitiation involving upstream ORFs regulates ATF4 mRNA translation in mammalian cells. *Proc Natl Acad Sci U S A* 101: 11269–11274.

109. Wang XZ, Lawson B, Brewer JW, Zinszner H, Sanjay A, Mi LJ, Boorstein R, Kreibich G, Hendershot LM, Ron D. 1996. Signals from the stressed endoplasmic reticulum induce C/EBP-homologous protein (CHOP/GADD153). *Mol Cell Biol* 16: 4273–4280.

110. Zinszner H, Kuroda M, Wang X, Batchvarova N, Lightfoot RT, Remotti H, Stevens JL, Ron D. 1998. CHOP is implicated in programmed cell death in

response to impaired function of the endoplasmic reticulum. *Genes Dev* 12: 982–995.

111. Oyadomari S, Harding HP, Zhang Y, Oyadomari M, Ron D. 2008. Dephosphorylation of translation initiation factor 2alpha enhances glucose tolerance and attenuates hepatosteatosis in mice. *Cell Metab* 7: 520–532.

112. Suragani RN, Liu S, Zhao W, Chen J-J. 2008. Oxidative stress signaling pathway of heme regulated eIF2α kinase in protection of β-thalassemia. *Blood* 112: 53.

113. Masuoka HC, Townes TM. 2002. Targeted disruption of the activating transcription factor 4 gene results in severe fetal anemia in mice. *Blood* 99: 736–745.

114. Coutts M, Cui K, Davis KL, Keutzer JC, Sytkowski AJ. 1999. Regulated expression and functional role of the transcription factor CHOP(GADD153) in erythroid growth and differentiation. *Blood* 93: 3369–3378.

115. Patterson AD, Hollander MC, Miller GF, Fornace AJ, Jr. 2006. Gadd34 requirement for normal hemoglobin synthesis. *Mol Cell Biol* 26: 1644–1653.

116. Jousse C, Oyadomari S, Novoa I, Lu P, Zhang Y, Harding HP, Ron D. 2003. Inhibition of a constitutive translation initiation factor 2alpha phosphatase, CReP, promotes survival of stressed cells. *J Cell Biol* 163: 767–775.

117. Uma S, Thulasiraman V, Matts RL. 1999. Dual role for Hsc70 in the biogenesis and regulation of the heme-regulated kinase of the α subunit of eukaryotic translation initiation factor 2. *Mol Cell Biol* 19: 5861–5871.

118. Uma S, Hartson SD, Chen J-J, Matts RL. 1997. Hsp90 is obligatory for the heme-regulated eIF-2α kinase to acquire and maintain an activatable conformation. *J Biol Chem* 272: 11648–11656.

Chapter 5

ROLE OF HEME IN BRAIN FUNCTIONS

Dr. Jekyll or Mr. Hyde?

Tatyana Chernova and Andrew G. Smith

Overview

The main role of heme in biological systems is usually viewed as that of a prosthetic group of the proteins involved in oxygen transport and sensing, metabolism, detoxification, and production of the signaling molecules carbon monoxide (CO) and nitric oxide (NO), thereby conferring upon them particular functional properties that vary according to the precise biological requirements. The turnover of heme and regulation of the supply for these functions is complex and involves a variety of physiological interactions of different cell types. More recently, it has become clear that heme, when it is not bound to a functional protein as a prosthetic group, can also serve as a regulatory molecule whose binding to specific targets alters their function. Findings have been especially interesting in the field of neurobiology. This chapter describes functions of heme in the brain and emerging knowledge of its role in normal and neuropathological processes. We give an outline of heme metabolism in the brain and present our views on contradictory roles of heme in neuronal cells. Neurotoxic and beneficial effects of heme are illustrated for neurological conditions, such as intracerebral and subarachnoid hemorrhages, neurodegenerative diseases and aging. Neuroprotecive roles of heme and the products of its degradation are reviewed here in the context of demand and availability of heme. We give an account of detrimental effects of compromised heme levels in neuronal cells and interaction of heme with neuronal proteins associated with neurodegenerative diseases. We describe emerging regulatory roles for heme in neurons and heme-driven signaling events.

Heme synthesis and disturbances pertinent to neuronal cells

The fundamentals of mammalian heme synthesis were first elucidated with rabbit reticulocytes and later studies progressed to liver and to a limited extent other organs (1). Nonerythroid heme synthesis is often assumed to be similar in all organs, but subtle and important differences probably exist as noted for the harderian gland of rodents (2). Of the eight steps specific to heme synthesis, the initial and last three take place in the mitochondria whereas the intermediate ones occur in the cytosol (Fig. 1).

The first precursor, 5-aminolevulinic acid (5-ALA) is formed by the condensation of glycine with succinyl CoA in the mitochondrion (1). In erythroid heme synthesis, which accounts for the majority of the body's heme production, regulation of the erythroid-specific 5-ALA synthase gene (*ALAS2*) occurs by iron responsive and other elements coupled to requirements for new hemoglobin. Although nonerythroid heme synthesis may account for much less heme production, it is vital to sustain many cellular systems in all tissues, not the least oxidative respiration in mitochondria. The expression of the nonerythroid specific enzyme (*ALAS1* gene) is also an important regulating step but markedly different from that of *ALAS2*. It has long been considered to be a consequence of a "regulatory" heme pool negatively controling transcription, mRNA stabilization and protein processing, although the exact mechanisms have still

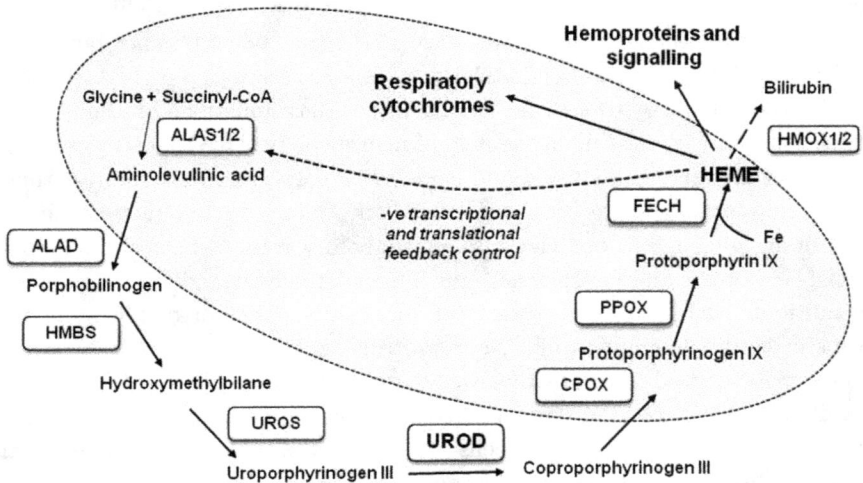

Fig. 1.

not been fully elucidated (3). As might be expected, a variety of factors have been implicated such as nutrition, hormones and induction of particular cytochrome P450 isoforms that not only act at the transcription level but also on ALAS protein processing and turnover.

Condensation of two molecules of 5-ALA by ALA dehydratase (*ALAD* gene) in the cytosol leads to the first cyclic product, the pyrrole porphobilinogen (1). Polymerization of four porphobilinogens creates a linear tetrapyrrole, 1-hydroxymethylbilane (catalysed by hydroxymethylbilane synthase, alias porphobilinogen deaminase; *HMBS* gene). The semistable bilane can cyclise chemically to the symmetrical macrocyclic tetrapyrrole uroporphyrinogen isomer I, which cannot be converted to heme, and is ultimately excreted. However, under normal conditions, most of the bilane is cyclized enzymatically (UROS) and undergoes spiro transformation, inverting one pyrrole ring, to form the asymmetric isomer uroporphyrinogen III (Fig. 1). Both uroporphyrinogens are highly unstable and in favorable conditions will rapidly oxidise by six electron withdrawals to the respective porphyrins and be lost to heme synthesis. Subsequently, four sequential decarboxylation steps of uroporphyrinogen III converts acetate side-chains to methyl groups (Fig. 1), catalysed by uroporphyrinogen decarboxylase (UROD) (1). At this point, following transport of the product coproporphyrinogen III back into mitochondria, two further decarboxylation steps occur by an oxidative mechanism of the adjacent propionic acid side-chains. The resulting divinyl product, protoporphyrinogen IX, although spontaneously oxidising to protoporphyrin IX, is oxidised enzymatically in a controlled manner by protoporphyrinogen IX oxidase (*PPOX* gene) (Fig. 1). Finally, heme is formed by insertion of Fe^{2+} into protoporphyrin IX to form heme by the action of a ferrochelatase (*FECH* gene) (1). Thus the process of heme formation from simple precursors starts and finishes in the mitochondria. The asymmetric macrocyclic molecule, contains two vinyl side chains on one side and two adjacent carboxy propyl side chains on the other. In different roles of heme, these allow variable covalent or hydrogen bonding to surrounding peptides and together with iron coordination results in exquisitely specific oxygen binding and oxidation mechanisms. For many functions besides respiratory cytochromes, such as in microsomal cytochrome P450s and nitric oxide synthase, heme must be exported from the mitochondria and shuttled across the cell to sites of utilization.

Phenotypic consequences of inherited defects in enzymes of the heme pathway in humans (porphyrias) are known for all genes (4). Most result in decreased activity but the recently reported mutation of *ALAS2* is a gain of function (5). Some porphyrias are expressed in heterozygotes and in others the phenotype is only apparent in patients homozygous for the mutation. Deficiency of ALAS2 in erythroid tissue does not lead to a porphyria but causes X-linked sideroblastic anemia with iron deposition in mitochondria (4). Although the porphyrias are relatively rare, the frequency varies greatly for each gene and population. For each porphyria, one gene mutation dominates but its effect may be dependent on expression of other genes or factors such as diet and hormones (6). The consequences of enzyme deficiencies of the heme synthesis mean that intermediates in the pathway, such as 5-ALA and porphyrinogens, may accumulate in tissues and/or are excreted in elevated amounts as well as being potentially toxic. In addition, the porphyrinogen intermediates are unstable, especially to light, and form the respective porphyrins. Porphyrias have been classified clinically as either hepatic or erythropoietic, depending on whether the primary gene affected is predominantly expressed in the liver or bone marrow, and as acute or cutaneous depending on symptoms (4). Broadly, in acute porphyrias there is an overproduction of simpler precursors such as 5-ALA and porphobilinogen before porphyrinogens are formed (or the oxidative porphyrin products). This can be manifested clinically by neurological disturbances involving mechanisms that have not yet been resolved completely but could involve neurological heme insufficiency or neurotoxicity of 5-ALA that has been produced in the liver.

The biosynthesis and control of heme in the brain *per se* has received relatively little systematic attention even though we now know that many critically important oxidative biotransformation and signaling processes occur there. The brain has a lower, but significant, heme content than the liver (Table 1) but has a higher metabolic rate and higher affinity for oxygen than liver tissue (7). It would be expected that the regulation of heme synthesis in neurological tissue is under complex and subtle control involving regulatory heme pools to balance internal and external demand (Fig. 1).

It has been known for many years that a range of drugs and chemicals, including metals, could disrupt heme synthesis at many stages including 5-ALA synthesis and protoporphyrinogen oxidase and ferrochelatase steps, and could be used as models to explore phenotypes of some human porphyrias (8–9). One of the simplest interactions is the well known inhibition of 5-ALA dehydratase after lead exposure (10). Elevated serum 5-ALA

Table 1. Comparison of heme concentrations in mouse brain fractions, kidney and liver.

Tissue	Heme content	
	pmol/mg tissue	*% of liver value*
Liver	7.09 ± 0.78[a]	100.0 ± 11.0
Kidney	2.68 ± 0.38	37.7 ± 5.3
Brain olfactory bulb	2.81 ± 0.13	39.6 ± 1.8
Brain cortex	2.35 ± 0.58	33.1 ± 8.2
Brain subcortical white matter	2.27 ± 0.03	32.0 ± 0.4

Values represent means (n = 3) ± S.E.M. and are taken from Ref. (1).

during prenatal development has been proposed as a risk factor for schizophrenia (11). In rats, brain ALAS activity changes markedly during early *postpartum* development (12). Another interesting example is succinyl acetone which is a powerful inhibitor of ALA dehydratase and produced in human type I tyrosinemia which is associated with neurological symptoms (13). ALA dehydratase activity is depressed *in vivo* in mouse brain by succinyl acetone administration (14) and the metabolite has been used in many *in vitro* studies to depress heme synthesis. Overall, brain ALAS1 activity in rodents is much less affected by drugs that cause its induction in rodent liver so that a deficiency of heme might account for neurological symptoms of lead poisoning and acute porphyria in circumstances of critical requirement. In aging animals there is probably a decrease in heme formation which could compound matters (15). Depression of heme availability for tryptophan pyrrolase (dioxygenase) was at one time proposed as a contribution to neurologic attacks in acute hepatic porphyrias (16). Injection of succinyl acetone or hematin directly into the brain has led to up- and down-regulation of ALAS activity respectively consistent with a heme pool regulating its own synthesis (17–18). Details of how heme and its catabolism are presently thought to play a role in brain function and in neuropathological development are discussed below.

Contradictory roles of heme as a toxic and a physiological molecule

Heme is an essential molecule with contradictory biological functions. The beneficial properties of heme universal to all living organisms include an

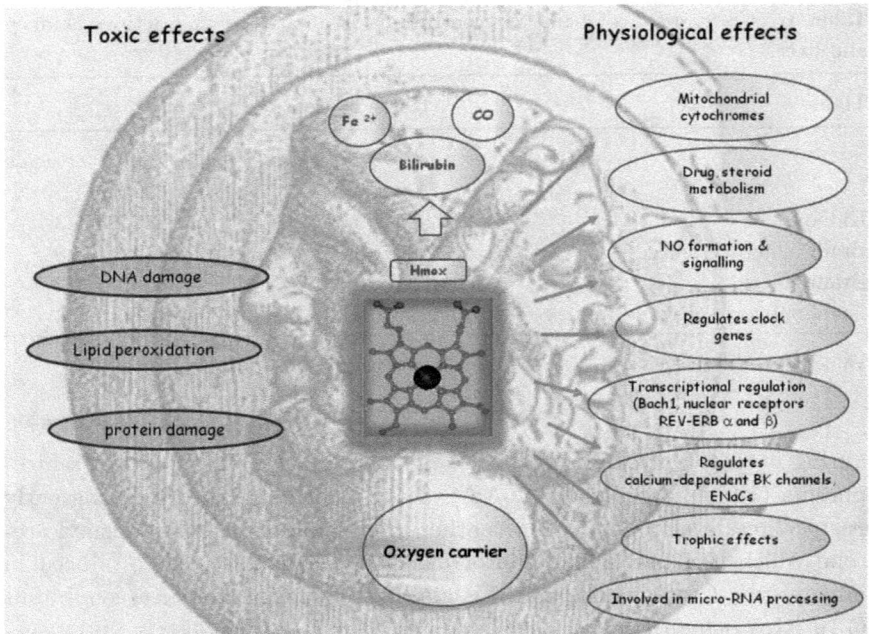

Fig. 2. Contradictory functions of heme in mammalian biology as a physiological and toxic molecule.

ability to bind oxygen, nitric oxide, and carbon monoxide, and the capacity to transfer electrons. At the same time, heme at high concentrations can promote free radical formation and lipid peroxidation, resulting in cell damage and tissue injury (19) (Fig. 2).

A fine balance between availability of heme for beneficial but not toxic functions is achieved by multiple levels in regulation of heme metabolism, storage and degradation. However, under certain pathological conditions, this equilibrium can be distorted.

Severe hemolysis occurring during pathological states, such as hemorrhagic strokes, intracerebral and subarachnoid hemorrhage (SAH) and ischemia reperfusion, results in high levels of free heme in brain tissues. Spontaneous intracerebral hemorrhage (ICH) accounts for 15%–20% of all strokes (20–21) and has a poor prognosis. While SAH represents only about 7% of all strokes, it is the most deadly — with more than a 50% fatality rate (22). Within hours after a CNS hemorrhage, hemoglobin (Hb) is released from extravasated erythrocytes into the extracellular space following cell lysis, mediated by the complement system (23). Free heme can also be liberated from other various intracellular and extracellular hemoproteins during cell lysis (24). Any free hemoglobin ($HbgFe^{2+}$) released into the tissues or circulation is promptly converted by spontaneous oxidation to methemoglobin ($HgbFe^{3+}$), and during deoxygenation, some of the oxygen leaves as a superoxide $O_2^{\cdot-}$ radical. Heme moieties that are the oxidized form of (porphyrin-Fe^{3+}), such as in methemoglobin, dissociate more readily from the protein, than the reduced heme that is tightly bound to the proteins such as in hemoglobin. (25–26). Hemin released from hemoglobin is present at high micromolar concentrations in intracranial hematomas (27–28). Release of "free" heme from methemoglobin into adjacent brain tissue was considered a factor that substantially contributes to the morbidity of patients with ICH (29).

The physiological mechanisms of removing free heme include heme oxygenase systems (HO-1, HO-2) and extra-HO systems (formation of non-toxic complexes with hemopexin (HPX) or albumin, scavenging free iron by ferritin or degradation by reduced glutathione (GSH), NADPH-cytochrome P-450 reductase etc.) (19). HPX is expressed in the brain and has a much higher affinity for heme ($K_d \sim pmol/L$) than does albumin and may function physiologically to regulate the balance between free heme and bound heme, and/or regulate heme degradation (30). HO-1 is normally expressed at a very low level, but is rapidly induced in astrocytes, microglia, and some neurons by extravascular hemoglobin, hemin, and a variety of oxidants (31). HO-2 is constitutively expressed, predominantly by neurons (32).

Neurotoxic effects of heme

Under circumstances of extracellular accumulation of extremely high amounts of free hemoproteins or when heme defensive mechanisms and detoxification systems are not sufficient or out of reach, heme exerts its damaging effects.

Acute damage

Heme released from intracellular hemoproteins is a major source of redox-active iron, and generation of reactive oxygen species (ROS) causes various modifications to DNA bases (33) and peroxidation of proteins and lipids. Free heme can catalyze the generation of oxygen radicals by the Fenton reaction, leading to oxidation of low-density lipoproteins (LDL) (34–35), which can be accompanied by covalent cross-linking of the LDL protein, apo B (36). Lipid peroxides can further react with redox metals (37).

The neurotoxic effects of hemoglobin were demonstrated on models where it was added to cultured neurons *in vitro* (38) or injected into the brain *in vivo* (39). Free heme (10 to 50 μmol/L) significantly reduced cell viability in primary cortical cultures, whereas the heme-HPX complex was not toxic to primary neurons (30). Goldstein and colleagues (28) showed that exposure to hemin at physiologically relevant concentrations for 24 hours resulted in the death of human neuron-like cells that progressively increased between 3 and 30 μM (EC_{50} approximately 10 μM); protoporphyrin IX, the iron-free congener of hemin, was not toxic. Cell death was preceded by a marked increase in cellular ROS. The cellular damage was predominantly necrotic and largely prevented by iron chelators. Accumulation of iron in the brain after ICH has a long lasting effect *in vivo* and can contribute to brain atrophy and prolonged neurological deficits post hemorrhage (40). Neurons and glial cells had differential response to hemoglobin exposure: the latter resulted in widespread and concentration-dependent neuronal death (EC_{50} at 1–2.5 μM) without injuring glia (41). This may reflect a greater vulnerability of neurons to the products of heme degradation and ROS. Despite the limitations of the cell culture approach, which does not allow a mitigation of hemin toxicity by endogenous antioxidants or by compounds that complex with it in the intact CNS, the studies provide convincing evidence of the deleterious effects of heme on neuronal cells, especially considering that concentrations of free heme in hematomas can reach high micromolar levels: for example, >350 μM was reported in experimental subarachnoid hematoma (42). An interesting *in vitro* model of heme neurotoxicity has been developed by Jaremko and colleagues (43) in which clotted blood was co-cultured on porous membrane inserts with cortical neurons and glia. The study demonstrated that hemoglobin release from the clot was accelerated by the presence of neurons and glial cells, neurons were selectively vulnerable to hemoglobin toxicity and protected by deferoxamine (DFO), antioxidants and the *N*-methyl-*D*-aspartate (NMDA) receptor antagonist MK-801. Excitotoxic and iron-dependent mechanisms contributed to the inherent neurotoxicity of free hemoglobin after intracranial hemorrhage (43).

Chronic damage

A significant amount of cell damage occurs in the weeks following hemolysis-related events, and this brain injury is associated with brain edema, disruption of blood–brain barrier (BBE) and development of chronic inflammation. Several mammalian models have been developed to study mechanisms of brain injury after ICH. Infusion of autologous blood into the cortex and caudate nucleus has been used extensively to study mechanisms of brain injury and hemolysis has been proven to be a decisive step in inducing cerebral vasospasm (44). Infiltration of the brain by blood components induced an inflammatory response including overexpression of endothelial and leukocyte adhesion molecules and recruitment of macrophages and neutrophils to the site of hemorrhage, followed by an increase of TNF-α release from circulating macrophages in *in vivo* models (44–45). Pathways leading to BBB dysfunction include neutrophil and macrophage recruitment, microglial activation, activation of the kinin system and mitochondrial alterations, with a causative factor being an overproduction of ROS (46). A deleterious role of metalloprotease (MMP)-9 in acute brain injury within the first 3 days after ICH has been reported (47). After activation of proteases by free radicals, matrix MMPs, particularly MMP-9 and -2, can digest the endothelial basal lamina leading to BBB opening (48). *In vitro* studies demonstrated that cultured neurons were killed by MMP-3, and that neuronal death was most marked when all three proteases, MMP-3, MMP-9 and thrombin, were combined in the treatment (49). An increase in caspase-9-like and caspase-3-like activity in apoptotic-like pathways demonstrated *in vitro* might have contributed to brain cell death after intracerebral hemorrhage (50). The pathways implicated in pathogenesis of immediate and delayed brain damage converge on the common mechanism, generation of ROS.

Neuroglobin, hemoglobin and their neuroprotective roles

In 2000, Burmester and colleagues (51) identified a novel hemoprotein, a member of the globin superfamily, which was found to be expressed predominantly in the brain and therefore was called neuroglobin. It is a monomeric globin that is distantly related to hemoglobin and myoglobin. In humans, neuroglobin is expressed at submicromolar concentrations in most cell types, but is found at high levels (approaching millimolar concentrations) in brain neurons, retinal cells and some endocrine cells and localized close to mitochondria (52). Neuroglobin expression is induced by neuronal hypoxia and cerebral ischemia (53) and the neuroprotective role of neuroglobin has been shown both *in vitro* and *in vivo* in models of hypoxic–ischemic brain injury. It has been suggested that human neuroglobin may be an efficient scavenger of reactive oxidizing species and thus may play a role in the cellular defence against oxidative stress (52). Subsequent work has shown that the affinity of neuroglobin for oxygen is significantly lower than that for myoglobin or haemoglobin, and under normal physiological conditions neuroglobin is expected to be only partly saturated with oxygen (54), therefore neuroglobin is unlikely to function as an oxygen delivery system. Further research by T. Brittain's and A. Fago's groups suggested that ferrous neuroglobin exerts its protective effects via modulation of the early events in the intrinsic apoptotic pathway by the rapid reduction of cytosolic ferric cytochrome *c* (55–56). The level of expression of neuroglobin in the penumbra area makes a dramatic difference in the level of neuroprotection (57). The molecular mechanisms by which neuroglobin exerts its protective effect from Alzheimer's (AD) disease are not entirely clear. The protective role of neuroglobin in AD is ascribed to suppression of β-amyloid–induced ROS/RNS overproduction and lipid peroxidation and reduction of β-amyloid-induced mitochondrial dysfunction in PC12 pheochromocytoma cells (58). Neuroglobin attenuates β–amyloid neurotoxicity in murine cortical neuronal cultures *in vitro* and the Alzheimer phenotype in transgenic models *in vivo* (59). Over-expression of neuroglobin in the neuroglobin transgenic mice resulted in a marked decrease in ROS/RNS production and lipid peroxidation after ischemia–reperfusion in the CA1 region of hippocampus. Neuronal injury was also markedly reduced, possibly via intrinsic antioxidant properties of neuroglobin (60).

Human ferric neuroglobin has been shown to act as a guanine nucleotide dissociation inhibitor for the alpha subunit of heterotrimeric G proteins. These findings suggest that human neuroglobin may function as a regulator of signal transduction in the brain and that the heme moiety is crucial for this redox-coupled sensing (61).

It has been reported recently that hemoglobin possesses neuroprotective qualities and is specifically expressed in neurons of the cortex, hippocampus, and cerebellum of the rodent brain, but not in astrocytes and oligodendrocytes; the pattern of its expression differs from the neuroglobin expression pattern on both cellular and subcellular levels (62). These data, however, partly contradict the findings of Biagioli and colleagues (63), who detected α- and β-chain transcripts of hemoglobin and protein (Hb-IR) in mouse brain in a subpopulation of dopaminergic neurons, cortical and hippocampal astrocytes, and all mature oligodendrocytes. *In vitro* experiments demonstrated that the protein expression does not fully overlap with the transcript distribution. The authors also showed that hemoglobin is expressed in the substantia nigra of the human post-mortem brain (63). It has been proposed that neuronal hemoglobin expression is associated with facilitated oxygen uptake in neurons, and hemoglobin might serve as an oxygen capacitator molecule (62). In invertebrates, neuronal hemoglobin may provide a homeostatic mechanism in hypoxic conditions, serving as intracellular storage molecule for oxygen (63). The hypoxia-induced growth factor erythropoietin (EPO) stimulates hemoglobin expression in the mouse brain (62). This, however, raises a question about a source of heme for hemoglobin in the brain. It is well documented that the first step in the heme biosynthetic pathway is conducted by two separate enzymes in non-erythroid and erythroid tissues, ALAS1 (nonspecific 5-aminolevulinate synthase) and ALAS2 (erythroid 5-aminolevulinate synthase), the latter regulating heme synthesis for hemoglobin. The expression of *ALAS2* in the brain has been reported absent (63) or extremely low (64) and there is always a risk of contamination of a brain tissue sample with the blood, which may account for *ALAS2* detection.

Heme oxygenases and neurodegenerative diseases: Where does heme come from?

Heme oxygenase (HO) enzymes catalyses the first and rate-limiting step in heme degradation to ferrous iron, carbon monoxide and biliverdin (Fig. 3). To date, three genes for isoforms of heme oxygenase have been identified: the inducible HO-1, and the constitutive HO-2 and HO-3. *HO-3a* and *HO-3b* genes were found to be processed pseudogenes derived from HO-2 mRNA and the isoform HO-3 protein has very low catalytic activity; its physiological functions probably include heme binding and oxygen sensing (65–66).

All the byproducts of HO activity play a significant role in physiological cell functions: biliverdin is further converted to a potent antioxidant, bilirubin (34), and carbon monoxide acts as a physiological regulator of cGMP and thereby may function as a neurotransmitter (67). Heme is the

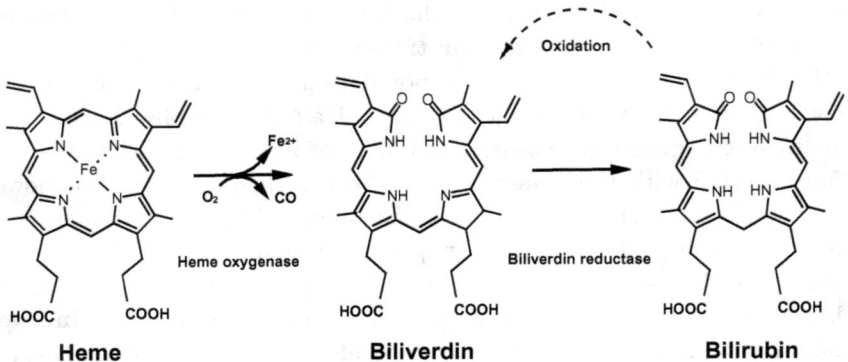

Fig. 3. In the first reaction, heme oxygenase cleaves the heme ring. The substrates for the reaction are heme, three molecules of oxygen and NADPH. The alpha-methylene group is released, the products are biliverdin, carbon monoxide (this is the only endogenous source of carbon monoxide), iron (II), and NADP$^+$. In the second reaction, biliverdin is converted to bilirubin by biliverdin reductase.

only source of carbon monoxide in the brain. Heme oxygenase-1 (HO-1), also known as heat shock protein (Hsp)-32, is a 32-kDa stress protein with relatively low levels of expression in the brain, but is highly inducible by numerous noxious stimuli in both neuronal and glial cells (68). HO-1 expression and activity in the brain have been intensively studied for its potential role in neuroprotection. HO-2 is a constitutively expressed 36-kDa protein which is abundant in the brain, and the majority of basal HO activity is attributed to the HO-2 isoform in neuronal tissues. HO-1 has been implicated in the pathogenesis of AD and other aging-related neurodegenerative disorders. In individuals with AD and mild cognitive impairment, immunoreactive HO-1 protein is overexpressed in neurons and astrocytes of the cerebral cortex and hippocampus relative to age-matched, cognitively intact controls and co-localizes to senile plaques, neurofibrillary tangles, and corpora amylacea (69). Additionally, the spatial distribution of HO-1 expression in AD brains was found by Takeda and colleagues to be identical to that of the pathogenic conformational changes of tau protein, the major component of the neurofibrillary tangles (70). Furthermore, HO-1 overexpression was concomitant with reduced tau expression and this was accompanied by inactivation of the mitogen-activated protein kinase (MAPK)-cascade (70). The authors suggest that the change in oxidation state not only induces HO-1 expression but may also play a role in the regulation of the tau gene.

In Parkinson disease (PD), HO-1 is markedly overexpressed in astrocytes of the substantia nigra and co-localizes with Lewy bodies, characteristic cytoplasmic inclusions present only in the PD specimens of affected dopaminergic neurons (69). PD and control dopaminergic neurons were readily distinguishable from each other in these preparations because Lewy bodies exhibited intense HO-1 staining in their peripheries. Furthermore, HO-1 immunoreactivity in the nigral neurons of PD specimens was generally more prominent than that observed in the controls (69). It is generally accepted that the elevated HO-1 levels represent a mechanism to restore redox homeostasis by producing antioxidants (biliverdin) and to reduce inflammatory response. Activation of HO-1 expression has been used to achieve neuroprotective and anti-inflammatory effects *in vivo* and *in vitro* models. Pharmacological modulation of HO-1 levels (within therapeutic limits) in the brain has shown promising results in models of AD and PD. Drugs or dietary products that cross the blood–brain barrier efficiently and can activate HO-1 expression *in vivo* are considered as neuroprotective and anti-inflammatory agents (71). For example, curcumin (1,7-bis [4-Hydroxy-3-methoxyphenyl]-1,6-heptadiene-3,5-dione), a natural phenolic agent, extracted from the rhizome of *Curcuma Longa*, strongly induced HO-1 expression and activity in rat astrocytes and in cultured hippocampal neurons (72). An alternative approach to increase HO-1 expression in the brain was used by Sakoda and colleagues in a model of brain cold injury; they modulated HO-1 levels via the transcription factor Bach1, which is known to repress transcription of HO-1; $Bach1^{-/-}$ mice showed significantly higher HO-1 mRNA expression than $Bach1^{+/+}$ mice in all brain sites studied, and higher induction of HO-1 was observed around damaged tissues after cold injury in $Bach1^{-/-}$ than $Bach1^{+/+}$ mice (73).

However, the induction of HO-1 activity would increase the requirement for the substrate, heme (3, 34). This raises the question: does higher protein expression mean increased HO-1 activity? If this is the case, what is the source of heme, the substrate for HO? It has been reported that lentiviral overexpression of HO-1 in primary cortical cultures did not result in heme depletion. The expression of *ALAS1*, the rate-controlling enzyme of heme synthesis, did not change in neurons with seven-fold increased levels of HO-1 protein and the cultures did not have a heme-deficient phenotype, reflecting that cellular heme was not decreased and therefore activation of heme synthesis did not occur (74). A double-edged sword nature of HO-1 modulation comes across in many studies. Schipper and colleagues (75) recently reported the neuroprotective effect of *reduced* HO-1 expression, and this is an example of the contradicting findings that emphasizes a complexity of modulation of the heme degradation pathway for therapeutic purposes. Their work suggested that targeted suppression of HO-1 hyperactivity in glial cells may prove to be a rational and effective neurotherapeutic intervention in AD and related neurodegenerative disorders. These opposing views on whether HO has to be activated or suppressed to achieve neuroprotection may not be mutually exclusive; the extent and duration of HO-1 induction may be critical. For a given pathological condition in the brain tissues, local redox status may vary, and excessive heme degradation may result in cytotoxic levels of ferrous iron and CO. Therefore, the balance between benefits of antioxidant biliverdin and potential oxidative effects of liberated iron would be determined by the tissue microenvironment (75). Nevertheless, in order to produce heme catabolites by HO enzymes, intracellular heme would need to be present at levels stoichiometrically equivalent to the levels of metabolites shown to be cytoprotective (76); therefore, the source of additional intracellular heme remains unclear.

HO-2 is constitutively expressed in neurons and represents the majority of native HO activity within the CNS. The modulation of HO-2 levels in the brain has also attracted a lot of attention in order to achieve neuroprotective effects in various scenarios. It has been reported that HO-2 gene deletion increased the vulnerability of astrocytes to hemin in cultures containing only astrocytes (77). However, in the model based on mixed neuron/astrocyte cultures and treatment with hemin at concentrations that did not injure astrocytes, the opposite was observed: HO-2 deletion attenuated hemin-induced ROS formation, reduced levels of oxidized proteins and increased neuronal survival in knockout cultures (77).

The vulnerability of cultured primary neurons incapable of HO-2 expression (HO-2 knockout neurons) to hemoglobin was compared with that of wildtype neurons by Rogers and colleagues. Knockout neurons were reported to be less vulnerable to hemoglobin toxicity and could cope with the increased ROS generation better than HO-2 +/+ neurons (78). The expression of HO-1, ferritin, and superoxide dismutase expression in HO-2 −/− cultures did not differ significantly from that observed in HO-2 +/+ cultures. Cellular glutathione levels were slightly higher in knockout cultures, and HO-2 knockout neurons were more vulnerable than wildtype cells to inorganic iron (78). In contrast to this *in vitro* observation, the *in vivo* work by Chang and colleagues (79) showed that brain tissue from injured HO-2 KO mice exhibited decreased ability to reduce oxidative stress, as monitored by a Fe^{2+}/ascorbic acid-mediated carbon monoxide generation assay. Interestingly, total HO activity was substantially reduced in HO-2 KO mice, despite induction of HO-1 in this model. The authors also demonstrated that HO-2 KO mice sustain increased cell loss, reduced motor recovery, and a greater susceptibility to lipid peroxidation after traumatic brain injury compared to HO-2 wildtype mice (79).

Emerging regulatory roles for heme in neurons

Heme is crucial for sensing and utilising oxygen in neurons; both the production of nitric oxide and some of its actions are mediated through hemoproteins, such as guanylyl cyclase (80). Recent studies showed that heme can modulate functions of selected proteins in acute mode via a mechanism of reversible binding to heme regulatory motifs. Examples of mammalian proteins regulated by direct reversible interaction with heme include transcription factors Bach1, nuclear receptors REV-ERBα and REV-ERBβ, HRI (eIF-2 alpha kinase), microRNA processing protein DiGeorge critical region-8 (DGCR8), ion channels (SloBK potassium channel and epithelial sodium channel ENaCs) (81–86) (Fig. 4).

Emerging roles for heme molecules, which are not associated with hemoproteins as a prosthetic group, have mapped a new area in the field of heme biology, the area where heme acts as a signaling molecule.

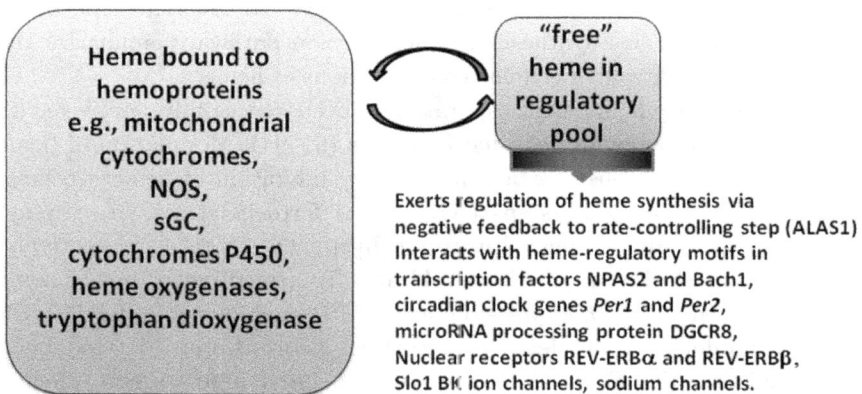

Heme bound to hemoproteins e.g., mitochondrial cytochromes, NOS, sGC, cytochromes P450, heme oxygenases, tryptophan dioxygenase

"free" heme in regulatory pool

Exerts regulation of heme synthesis via negative feedback to rate-controlling step (ALAS1)
Interacts with heme-regulatory motifs in transcription factors NPAS2 and Bach1, circadian clock genes *Per1* and *Per2*, microRNA processing protein DGCR8, Nuclear receptors REV-ERBα and REV-ERBβ, Slo1 BK ion channels, sodium channels.

Fig. 4.   Interrelationship between heme pools and functions.

A small, so far unspecified, proportion of "undesignated" or "free" heme (i.e. not associated with a specific functional role in a protein) forms a "pool" which can act in a signaling mode. It is not clear in what form heme exists in this pool, possibly bound to chaperone or transporter proteins, but it is clear that this "free" heme can regulate complex biological events. Free heme exists in cells at very minute concentrations, estimated as 100 nM (3). The regulatory functions of free heme include the repression of *ALAS1*, the rate-limiting enzyme in heme biosynthesis, and the induction of microsomal heme oxygenase-1 (HO-1) (3). In other words, heme determines its own fate. The regulatory mechanisms for heme synthesis, storage and degradation are very tight and provide a state of cellular equilibrium, whereby heme is readily available for its functions but protected from taking part in formation of cytotoxic ROS. Transcriptional regulation of *ALAS1* is tissue specific and in the brain initiation of transcription by drugs is influenced by the factors acting within and upstream of TATA box, in contrast to, for example, the liver. The transcription in liver cells is TATA box-driven (87) and stimulation of *ALAS-1* by drugs is less marked in the brain than in the liver, which may have implications in the assessment of neuroactive drugs and future drug design (88). Free heme also regulates *ALAS1* gene expression by reducing mRNA stability (89). Newly synthesized ALAS1 precursor protein has to be translocated to mitochondria, and this process is also regulated by heme (90). The catabolism of heme is regulated at the level of HO-1, the rate-limiting inducible enzyme. Regulation of gene expression for HO-1 occurs by way of free heme-mediated derepression of Bach1, a mammalian heme-responsive transcription factor that suppresses the activation of the HO-1 gene (53). Thus, multiple levels of regulation in heme synthesis and degradation pathways emphasize the importance of controlling cellular concentrations of heme.

Manipulating heme synthesis rate has been successfully employed to study effects of heme on neuronal functions (91–94). Modulation of heme levels in cultured mouse cortical neurons by inhibition of heme synthesis at two different enzymic steps (ALAD and ferrochelatese), by succinyl acetone (SA) and N-methylprotoporphyrin IX (NMP) respectively, resulted in marked up-regulation of *ALAS1*. Treatment with 100 nM heme daily restored *ALAS1* expression and culture phenotype (92–93). In another model of heme-deficiency, cortical neurons were derived from heme-deficient the $Fech^{m1Pas}$ mutant (abbreviated here to *Fech* mouse) containing a point mutation in the ferrochelatase gene (95–96). Cultures displayed a characteristic *ALAS1* up-regulation compared to wildtype neurons (92). Interestingly, HO-1 mRNA and protein levels in heme-deficient cortical neurons were increased if heme deficiency persisted over 72 hours and could be a response to a changing redox state and accumulation of ROS (93). However, induction of HO-1 in these models is unlikely to contribute to depletion of heme because activation of HO-1 by heme is attributed to the levels >1 µM (34), and kinetics of heme association with HO-1 are reported as K_m 3 ± 1.61 µM, K_d 0.84 ± 0.21 µM in humans (97). In cultures with inhibited heme synthesis, free heme concentration must be significantly lower than 1 µM if 100 nM was enough to restore gene expression pattern and phenotype (93).

Recent studies revealed even more exciting regulatory roles for heme in neurons and demonstrated that heme is involved in the regulation of neuron-specific genes expression (92–94, 98). Zhu and colleagues studied the role of heme in nerve growth factor (NGF) signaling and showed that in PC12 cells, heme deficiency interferes with the Ras-mitogen-activated protein kinase (MAPK) pathway. That resulted in down-regulation of neuronal genes associated with NGF signaling in specific and selective manners. Inhibition of heme synthesis by SA caused a marked decrease in extracellular signal-regulated kinases (ERK1/2) activation by NGF and subsequent suppression of the ERK1/2 downstream effector, transcription factor CREB (98). Microarray analysis demonstrated an altered pattern of gene expression in neuronal cells, including genes for neurotransmitter receptors, synaptic vesicles proteins, cytoskeletal and trafficking proteins (98). Further work by this group has shown that the effects of heme deficiency on neuronal gene expression can be reversed by treatment with exogenous heme (94). The authors also demonstrated activation of the pro-apoptotic JNK signaling pathway concomitant with inadequate activation of the pro-survival ERK1/2 pathway and increased rate of apoptotic death in heme-deficient PC12 cells stimulated with NGF (94). In primary cultures of mouse cortical neurons, heme deficiency induced by inhibition of heme synthesis caused morphological changes similar to neurodegeneration (93). The progressive neurite fragmentation and disruption in the axonal/dendritic microtubule structure without an increase in cell soma loss caused by chronic heme depletion was similar to the early progress of many neurodegenerative diseases and age-related dementias (Fig. 5). Loss of projections network was followed, eventually, by cell death at late stage. The mechanism of neuronal degeneration involved impaired ERK1/2 phosphorylation and the rescue effect of heme was ascribed to the recovery of ERK1/2 sustained activation.

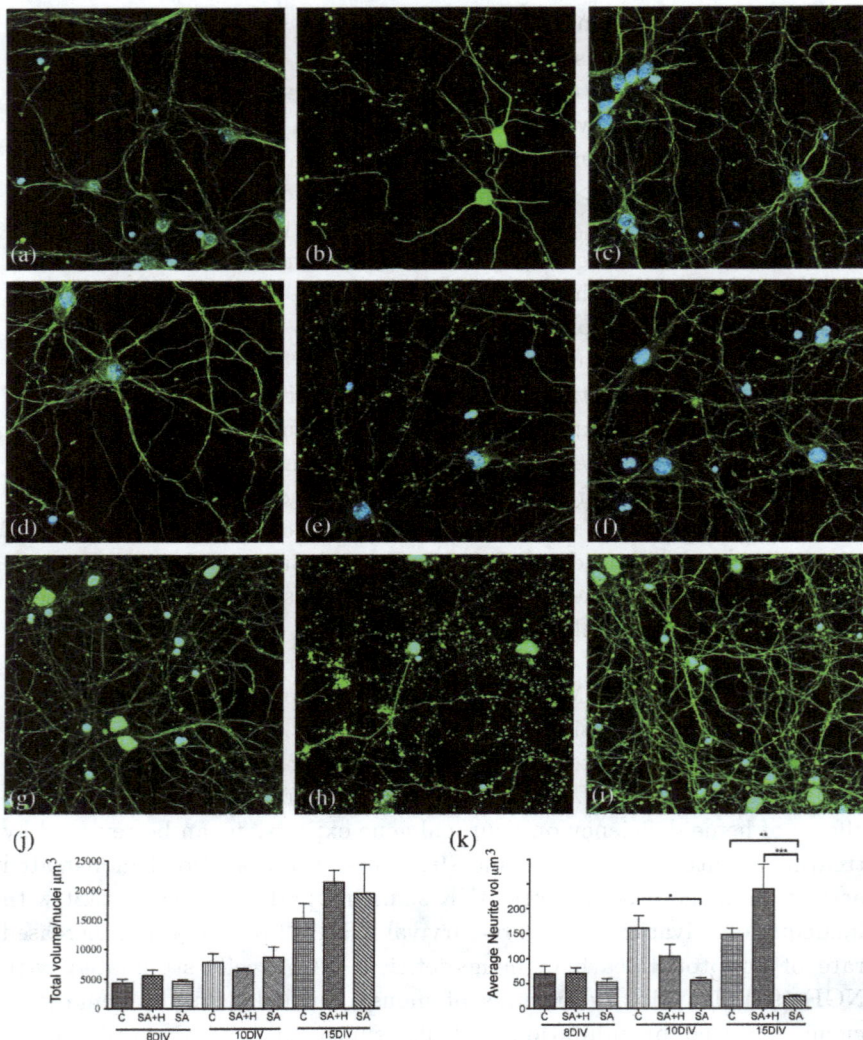

Fig. 5. Neurodegeneration of cultured cortical neurons caused by heme deficiency. Confocal micrographs of maximum projection Z series of cultured neurons stained for βIII-tubulin on DIV 8, 10 and 15. (a), (d) and (g) control cultures, (b), (e) and (h) cells treated with SA, (c), (f) and (i), cells treated with SA and heme. Quantification of immunostaining for βIII Tubulin reflecting the degree of neurodegeneration. (k) Average neurite volume in control and treated cultures at each time-point. (j) Total volume per nuclei in image fields examined. * indicates $p < 0.05$, ** $p < 0.01$ and *** $p < 0.001$. (For reference see (93)).

Suppression of the pro-survival ERK1/2 pathway and decreased activation of CREB was similar to that reported by Zhu and colleagues (98), although no activation of pro-apoptotic pathways was found in primary neurons (93). Similar results were observed in both inhibitor (SA-treated neurons) and mutant (*Fech^{m1Pas}* mouse neurons) models of heme deficiency and directly linked the phenomenon of ERK1/2 suppression with reduced heme synthesis (93). Further dissection of the corrupted molecular signaling revealed that failure of ERK1/2 sustained activation in heme-deficient neurons was NMDA receptor-dependent. NMDA receptors play a critical role in synaptic plasticity, learning and memory, and loss of synaptic NMDA receptors were correlated with an early functional decline in neurodegenerative diseases (99). Chronic heme deficiency caused reduction in expression, phosphorylation and function of NMDA receptor in cultured neurons. NMDA receptor-mediated and voltage-gated sodium currents were diminished but rescued in the cultures treated with exogenous heme (93). Receptors containing the NR2B subunit were found to play a predominant role in the rescue effect of heme on ERK1/2 phosphorylation in this model (93). Compromised heme availability was suggested to cause NMDA receptors dysfunction, leading to decreased Ca^{+2} influx, diminishing Ca^{+2}-dependent ERK1/2 activation, followed by loss of connectivity of neuronal processes, loss of synaptic NMDA receptors, further decreased Ca^{+2} influx, and so on, creating a vicious circle.

Further work by the authors (Chernova *et al.*, submitted) demonstrated that in cultured cortical neurons, heme can rapidly regulate NMDA receptor function. Two-hour treatments of heme-deficient culture with exogenous heme at a physiological concentration (100 nM) produced a marked increase of NMDA-evoked currents. However, if cultures were cultured pre-treated with a selective inhibitor of Src kinases PP2 [4-amino-5-(4-chlorophenyl)-7-(*t*-butyl) pyrazolo[3,4-*d*]pyrimidine], exogenous heme failed to increase NMDA-evoked current (Fig. 6a). Furthermore, exposure of a heme-deficient neuron to heme during patch-clamping (heme was added to the bath solution or to the patching solution inside of the pipette) resulted in a significant increase of NMDA-evoked current within minutes, indicating a signaling mode of action.

These data suggest an acute regulation of NMDA receptor by heme. The mechanism of this regulation is not entirely understood, although it is clear that tyrosine phosphorylation of the NR2B subunit plays a major role. The rapid rescue effect of heme on the receptor function may be

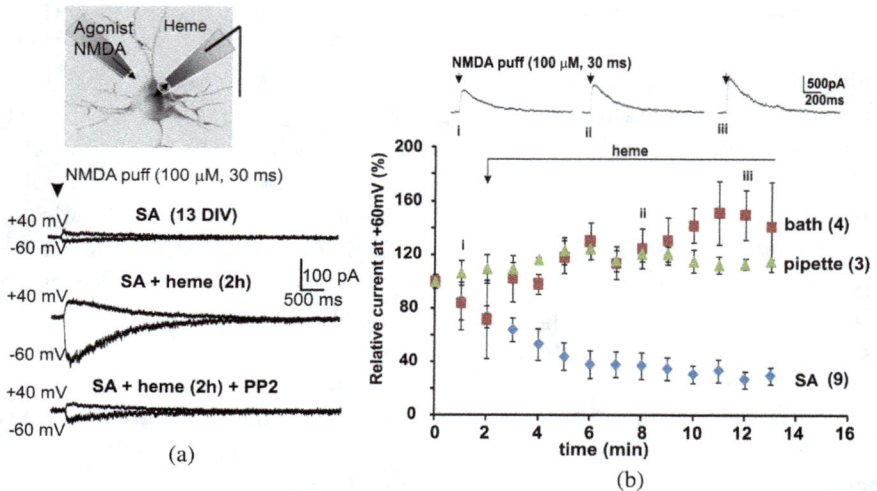

Fig. 6. Heme depletion suppresses NMDA-evoked currents. Incubation with heme reverses this change but not in the presence of Src family kinases inhibitor PP2. (a) Whole-cell currents evoked (representative traces) by pressure application of NMDA (indicated by arrow; 100 μM, 20 PSI, 30 ms) to cortical neurons at holding potentials of −60 and +40 mV. SA treated for 24 hours neurons (top trace), neurons treated with SA for 24 hours followed by incubation with heme for 2 hours in the absence (middle) or presence (bottom trace) of Src kinase inhibitor PP2. (b) Neurons treated with SA for 24 hours showed a 70% reduction in NMDA-evoked currents (diamonds) over a recording time of 13 min. When heme (0.1 μM) was applied in the bath at 2 min (indicated by the arrow) the initial current decline (similar to SA-treated, up to 2 min) stopped and then NMDA-evoked currents were potentiated by 40% over 13 min (squares). Top raw traces indicate representative currents at indicated time points (i–iii). When heme was presented in the patch pipette, currents started to increase immediately after patching (triangles).

exerted via direct interaction with proteins within the NMDAR complex, possibly those facilitating phosphorylation of the NR2B subunit by Src kinase (Chernova *et al.*, submitted).

Neurotrophic effects of heme

Neurotrophic effects of heme were first reported in 1978 by Ishi and Maniatis (100), who observed that hemin exposure increased the growth rate of mouse neuroblastoma cells in culture. Cells were incubated in

Table 2. Ability of compounds structurally or metabolically related to hemin to promote neurite outgrowth and effect of KCN, NaN$_3$ and NaF on response to hemin in cultured mouse neuroblastoma cells.

Treatment	Neurite outgrowth (%)*
a, Control	4.3±1.1
Hemin (10^{-4} M)	42.1±1.9
Biliverdin (10^{-4} M)	9.0±1.4
Porphobilinogen (10^{-4} M)	6.3±1.1
Chlorophyllin (5×10^{-5} M)	15.9±2.1
Chlorophyllin (10^{-4} M)	24.6±1.7
Chlorophyllin (2×10^{-4} M)	30.9±1.1
Pyrrole (10^{-4} M)	4.1±0.4
b, Control	2.7±0.2
Hemin (10^{-4} M)	43.0±2.2
KCN (5×10^{-4} M)	10.7±1.1
NaN$_3$ (5×10^{-4} M)	11.3±1.3
NaF (5×10^{-4} M)	10.9±0.8
Hemin (10^{-4} M) plus KCN (5×10^{-4} M)	40.1±1.8
Hemin (10^{-4} M) plus NaN$_3$ (5×10^{-4} M)	39.1±2.4
Hemin (10^{-4} M) plus NaF (5×10^{-4} M)	33.4±1.4

Cells were incubated for 3 hours in DMEMS medium with addition as indicated. Control cultures in (a) received 0.02% ethanol, which served as a solvent for compounds used in that experiment. The proportion of neurite-bearing cells in the cultures was determined.
* Values are means ± s.e.m. ($n = 4$). Data retyped from reference (100).

DMEM medium in the presence or absence of 10^{-4} M hemin for 3 hours after which the proportion of neurite-bearing cells was estimated. Hemin treatment resulted in the increase of the cells with neurites by 22%. The authors examined effects of other compounds related to hemin and the data are presented in Table 2 (100).

The authors suggested that the mechanism of rapid and reversible induction of neurite outgrowth may be associated with an ability of hemin to modulate activity of protein kinases, described previously in work by Greengard (101).

Nearly 30 years later, work by Shinjyo and Kita (102) demonstrated up-regulation of heme biosynthesis during differentiation of Neuro2a cells. A gradual increase in the cellular heme b level was observed between 12 and 48 hours, and an increase was evident after 72 hours. In contrast, the heme *a* content did not change significantly during retinoic acid-induced differentiation. The level of mitochondrial heme *c* (constituent of cytochrome *c*) was increased during differentiation. The levels of *ALAS-1* mRNA were increased during neuronal differentiation. The up-regulation of *ALAS1* expression was suggested to be related to the enhanced mRNA stability in a positive feedback response to the lower level of free heme (heme *b*), that was observed immediately after the induction of differentiation (102). A subsequent study (103) showed that this up-regulation of heme biosynthesis during differentiation is ROS-dependent. Furthermore, it was found that ROS-dependent induction of heme oxygenase, which degrades heme and acts as an antioxidant, and catalase, another antioxidant enzyme that contains heme as a prosthetic group, occurs during differentiation. Based on these observations, the authors suggested that increased heme biosynthesis following the degradation of heme protects Neuro2a cells from oxidative stress caused by ROS during differentiation. *In vitro* work showed that the differentiation of neuroblastoma SHSY5Y cells generated by nerve growth factor treatment was compromised by chemical-induced heme deficiency: heme-deficient cells failed to complete differentiation, lost their axons, and died within 3–4 hours after the differentiation induction (104). A partial explanation of this effect suggested by the authors was a corruption of signaling pathways as a result of compromised heme responsive motifs (HRM)-dependent factors, APP, and decayed mitochondria (104).

The effect of heme on neurites outgrowth was monitored in mouse primary cortical cultures (Fig. 7), and two-fold increase of projections volume in neurons treated daily with 100 nM hemin was found, however the mechanism is yet to be studied (Chernova *et al.*, unpublished).

Recently, Kimura and colleagues identified a novel secreted neurotrophic protein neudesin (105). Mouse neudesin mRNA was expressed in neurons but not in glial cells and was abundant in the developing brain. The neurotrophic activity exhibited in cultured mouse neurons by this protein was associated by the authors with activation of the mitogen-activated protein (MAP) and the phosphatidylinositol-3 (PI-3) kinase pathways. Their further work (106) showed that neudesin possesses a predicted cytochrome b_5-like heme/steroid-binding domain in its primary structure and that hemin binds to this domain of the neudesin molecule *in vitro* in Neuro2a cell. Association with hemin enhanced neurotrophic activity of the protein in mouse primary cortical neurons; cells cultured in the presence of neudesin-hemin had a greater neuronal survival and a lower proportion of

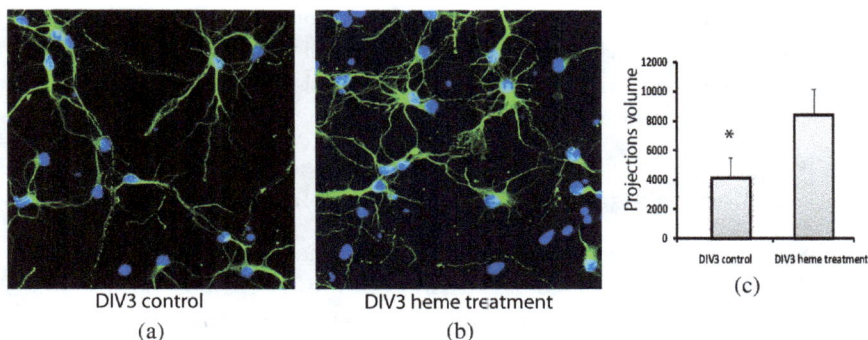

DIV3 control
(a)

DIV3 heme treatment
(b)

(c)

Fig. 7. Effect of heme on neuritis outgrowth in primary cultures of mouse cortical neurons. Cortical neurons were isolated from BALB/c embryos on E16 as described previously (92); 1 hour after plating, the growth medium was supplemented with 100-nM heme, followed by daily treatment with 100 nM heme. At DIV3 cultures were immunostained using anti-βIII-tubulin 1:3000 antibody (Cell Signaling Technology, Beverly, MA, USA) and secondary antibody (Alexa Fluor 488 at 1:500), followed by nuclear staining with 300 nM 4'-6-diamidino-2-phenylindole (DAPI). Images of cultured cells were obtained with a Zeiss LSM 510 META confocal microscope equipped with a 40x oil immersion lens. Images were collected as series of Z sections (approx. 0.5 micron) and reconstructed using Volocity software (Improvision, Coventry, UK) on three dimensional reconstructions of datasets using an intensity classifier and size exclusion criteria to identify neurites. Images were collected from at least five different randomly selected areas from each of two duplicate cell cultures. Average projections volume per nuclei in (a) control and (b) treated cultures was (c) compared and statistical significance was estimated using two-tailed student's t-test. * statistically significant, $p < 0.05$.

apoptotic death compared to controls. Neudesin-hemin promoted activation of the major pro-survival ERK1/2 pathway. Additionally, Fe(III)- protoporphyrin IX was reported essential for the neurotrophic activity of neudesin, but not Fe(II)- protoporphyrin IX or protoporphyrin IX. This report presenting the first extracellular heme-binding protein neudesin involved in intercellular signal transduction was followed by identification of neuferricin, a second novel extracellular heme-binding protein with a cytochrome $b5$-like heme/steroid-binding domain (107). Mouse neuferricin consisting of 263 amino acids has a heme-binding domain similar to neudesin and exhibits a characteristic shift of the Soret band from 402 to 422 nm under reducing conditions. Interestingly, in contrast to neudesin, neuferricin was shown to suppress cell survival in Neuro2a but stimulated neurogenesis in primary mouse neural precursor cells. The authors suggested that neuferricin is expressed in a region of undifferentiated neuronal cells, and that neuferricin promotes neurogenesis by suppressing the self-renewal of undifferentiated cells (107).

Heme can act as a gas-sensing regulator of brain functions

Heme-containing proteins can reversibly bind diatomic gaseous molecules, O_2, NO and CO (108) and can thereby serve as regulators of cellular responses to fluctuating concentrations of oxygen, carbon monoxide, and nitric oxide levels. If a regulatory heme-binding domain is coupled to a neighboring functional domain, a protein becomes a heme-based sensor. Ligand (O_2, NO and CO) binding by the sensor (heme) domain modulates the physiological role of the protein, such as DNA binding in the case of transcriptional factors or the catalytic reaction rate in the case of enzymes (see (109) for more details).

Work by Dioum and colleagues provided evidence that the mammalian neuronal transcription factor NPAS2, which is widely expressed in mammalian forebrain, binds heme to form a gas-regulated sensor (110). Both PAS domains of NPAS2 bind heme, and formation of a DNA-binding complex by NPAS2 with its partner BMAL1 is directly coupled to the heme status (110). The authors demonstrated that formation of the DNA-binding heterodimeric complex was inhibited by addition of CO at low μM concentration. It has been proposed that CO may function in the brain as a neurotransmitter and that NPAS2 may be its receptor (111). Carbon monoxide is a plausible candidate for a native ligand of NPAS2, although sensing of other heme ligands by this protein has not been ruled out (reviewed in (109)). The signaling function of carbon monoxide as a neurotransmitter/neuromodulator (80, 112) is utterly dependent on heme.

Heme and brain aging: Role of heme for protection from drug-induced neurotoxicity

Heme synthesis declines with age, and reduced ALAS1 activity together with increased activity of microsomal HO were suggested to be causative factors for the decreased microsomal concentration of heme (113). Compromised heme availability in the brain may have an effect on functions of hemoproteins including cytochromes P450 (CYP450), a superfamily of heme-containing proteins with a critical role in drugs and xenobiotics metabolism (114). Many drugs and chemicals affect heme synthesis, and this also occurs in the brain (17). The formation of CYP450 requires *de novo* synthesis of apoprotein and heme (115), therefore reduced heme synthesis may become a limiting factor in CYP450 availability. It has been suggested that the fundamental mechanism of drug induction is the same for cytochromes P450 and ALAS1 (116). The authors identified and characterized two enhancer elements which respond to prototypic inducer drugs and interact with the human pregnane X receptor NR1I2 and the human constitutive androstane receptor NR1I3 in 5-UTR of ALAS1. These findings show that up-regulation of cytochrome P450 apoprotein production and induction of heme synthesis are coordinated via transcriptional activation of ALAS1 gene (116–117). However, it's not clear if this coordination is distorted by aging. Although reduction of heme synthesis has been shown *in vivo* and *in vitro* (92, 113), studies of age-related impairment of cytochromes P450 activity in humans demonstrated highly controversial results (reviewed in (118)). CYP450 isoforms have been found in mammalian brains by various methods and mapped to different regions of the brain as well as to different subcellular locations (119). Much less is known about CYP450 in human brain; only CYP2D6 distribution throughout the human brain has been described (120). Induction of brain CYPs is brain region-, cell type-, isoform- and inducer-specific (119). The specific pattern of expression indicates that drug and metabolite levels are not evenly distributed throughout the brain, creating a scenario of differential cell/region vulnerability to neurotoxicity. Endogenous modulation of brain CYP450 by neurotransmitters was suggested by Gervasini and colleagues (121). However, this area remains not well understood, and the matter is not helped by the fact that the widely used plasma drug monitoring does not reflect drug kinetics in the brain. In addition, little is known about the role of these enzymes in age-related and xenobiotic-induced neurotoxicity. Older people use on average two to five prescription medications on a regular basis, and polypharmacy, defined as the use of five or more medications, occurs in 20% to 40% of this age group (122); this could increase the demand for synthesized *de novo* heme to provide an adequate amount for inducible CYP450 in order to metabolize prescribed drugs efficiently. Effects of compromised availability of heme on CYP 450 function was studied by Meyer and colleagues (114, 123). Examination of cellular subfractions showed that the highest levels of CYP1A1 was detected in cytosol in the brain tissue, whereas in the liver, CYP1A1 was exclusively localized in microsomes (114) where it is enzymatically active. The availability of heme in the brain was significantly lower than in the liver and authors suggested that reduced availability of heme is a limiting factor for incorporation of CYP1A1 into endoplasmic reticulum (ER) of the brain tissues *in vitro* (114). Transient expression of CYP1A1in COS-1 cells, combined with inhibition of heme synthesis by SA, provided additional evidence of impaired incorporation of CYP1A1 into ER under conditions of heme deficiency (114). This report was followed by *in vivo* work, which showed cytosolic persistence of CYP1A1 in heme-deficient transgenic mouse model with chronic impairment of heme synthesis, porphobilinogen deaminase-deficient (PBGD−/−) mice. Heme content was diminished in the liver and brain of PBGD−/− mice compared to wildtypes, and CYP1A1 protein was detected not only at the endoplasmic reticulum (ER), but also in the cytosol of PBGD−/− mice. CYP1A1 metabolic activity was restored *in vitro* in cytosol of PBGD−/− mouse brain tissue by heme treatment *in vitro* (123). The authors discuss the relevance of heme as a limiting factor to the functions of other members of the CYP450 superfamily, that in combination with age-related reduction in heme synthesis levels (113) highlights an important role of heme for protection from drug-induced neurotoxicity.

Recently a connection between the activity of CYP450 and memory was demonstrated for the CYP7B isoform in rats (124). CYP7B is highly expressed in the brain, particularly in the hippocampus and catalyzes the synthesis of neurosteroids (125). Work by Yau and colleagues examined whether hippocampal CYP7B bioactivity was related to spatial memory impairments with aging in rats. The authors showed selectively decreased CYP7B bioactivity in hippocampal tissue from cognitively impaired 24-month-old aged rats compared to age-matched cognitively intact rats and young 6-month-old controls. Furthermore, the cognitive impairment was ameliorated by treatment with the CYP7B product 7-alpha-hydroxypregnenolone; treated animals demonstrated improvements spatial memory retention (124).

May sufficient availability of heme delay aging?

Heme deficiency was suggested to be a factor in the mitochondrial and neuronal decay of aging because it causes a selective decrease of mitochondrial complex IV *in vitro* in human brain cell lines, leading to oxidative stress and corruption of Ca^{2+} homeostasis (104). Effects of heme on neuronal aging was demonstrated *in vitro* in mouse primary cortical cultures (92). An increase in the proportion of senescent cells in aging neuronal cultures was accompanied by the up-regulation of the genes for ALAS1 and HO-1, suggesting a state of relative heme deficiency. Furthermore, in neuronal models of heme deficiency, based on chemical inhibition of heme synthesis by SA or NMP, as well as in heme-deficient *Fech* neuronal cultures, senescent phenotypes developed faster. The pattern of neuronal gene expression in senescent neurons was found similar to that in heme-deficient cultures (92). The decrease of NMDA receptor subunits in an age-dependent manner *in vivo* is well established (126), and in cortical neurons, decreased expression of NMDA receptor subunits ζ1 and ε2 (NR1 and NR2B) was also associated with aging. Heme-deficient cultures exhibited premature decline in expression of NMDA receptor subunits and neurofilament light peptide (NF-L), and this effect was prevented by treatment with exogenous heme (92). These data associate heme deficiency with premature neuronal decay and implicate the role of heme metabolism in physiological and pathological aging.

Deficiency of heme and Alzheimer's disease

Alzheimer's disease (AD) is an age-related neurodegenerative disorder characterized by synaptic loss and neuronal cell death and manifested clinically by progressive cognitive dysfunction and memory loss (127). The pathogenesis of the disease is very complex and not entirely understood; many factors, such as abnormal metabolism of amyloid precursor protein (AAP), pathological accumulation of tau, oxidative stress, proinflammatory changes, mitochondrial decay and energy hypometabolism were implicated in AD development. The pathological hallmarks of AD are aggregations of amyloid-β peptide, formation of neurofibrillary tangles and selective synaptic and neuronal loss in the brain regions involved in learning and memory. Because of its high rate of oxygen consumption and its high content of polyunsaturated fatty acids, the brain exhibits increased vulnerability to oxidative stress. Elevated lipid peroxidation in the brains of AD patients reflects induction of oxidative stress (128), which then exerts secondary effects such as protein modification, oxidation and conformation changes (129). Disruption in the metabolism of iron has been postulated to play a role in the pathogenesis of AD, and iron-induced oxidative stress has been suggested a key factor in AD pathophysiology (reviewed by Altamura and Muckenthaler (130)). Heme is a common factor linking several perturbations in AD, including those associated with iron metabolism, mitochondrial complex IV, heme oxygenase, and bilirubin. Atamna and colleagues showed that chemical-induced heme deficiency results in a selective decrease in mitochondrial complex IV that leads to oxidative stress in neuronal cell lines and primary hippocampal neurons (104) and suggested that functional deficiency of heme, and the shortage of heme-a particularly (91), causes mitochondrial and neuronal decay of aging.

The metabolic consequences resulting from heme deficiency seemed similar to that in dysfunctional neurons in patients with AD and following work demonstrated heme binding to amyloid-β (131–132). Interaction of heme with amyloid-β *in vitro* was examined further and the binding was confirmed by spectral analysis (132). This work outlined a possible mechanism of heme deficiency, i.e. binding of heme to amyloid-β causes depletion of regulatory heme (132). Furthermore, differential affinity of heme binding to human and rodent amyloid-β peptide was reported and the authors suggested that this explains a susceptibility of humans to AD (Fig. 8) (131). Sequestration of heme in these complexes may result in functional heme deficiency and cause mitochondrial dysfunction.

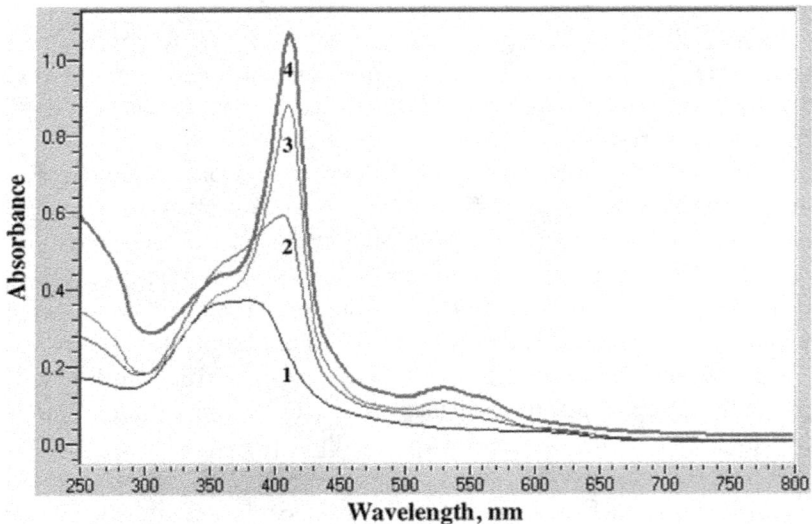

Fig. 8. Changes to the spectrum of heme induced by human amyloid-β peptides (huAb). The absorbance spectrum of heme (1) exhibits a red-shift and sharp increase in the Soret band at 412 nm, following binding with increasing concentrations of huAb. An additional absorbance peak appears at 530 nm. (1) 10-μM heme; (2) 10-μM heme + 10-μM huAb; (3) 10-μM heme + 30-μM huAb; (4) 10-μM heme + 60-μM huAb (with the permission of the authors, see (131) for reference).

The finding that heme-*a*/heme-*b* ratio was decreased in AD brain (132) also supported the hypothesis about role of heme deficiency in AD. Heme deficiency also induced formation of dimers and aggregates of APP in human brain cells and in rat hippocampal primary neurons (104). Homodimerization of amyloid precursor protein and amyloid-β has been reported previously as a factor implicated in the amyloidogenic pathway of AD (133). Howlett and colleagues reported the inhibitory effect of heme on amyloid-β (1–40 and 1–42) aggregation as well as amelioration of amyloid-β aggregation-dependent cell toxicity *in vitro* (134). On the other hand, the heme-amyloid-β complex has a peroxidise activity *in vitro*, that may contribute to neurotoxicity in cytopathology of the disease. Gatta and colleagues employed a model of heme deficiency to examine the effects of heme on processing of APP, where they inhibited ferrochelatase (FECH), the enzyme catalysing the terminal step in the biosynthesis of heme, converting protoporphyrin IX into heme. Inhibition of heme biosynthesis chemically (10-μM *N*-methylprotoporphyrin IX, NMP) or by FECH siRNA caused mitochondrial failure and increased oxidative stress in neuronal cells and altered the state and proteolytic processing of APP in cell line overexpressing the protein; the amount of secreted APP alpha was also reduced in heme-deficient cells (135). The authors also reported a difference between the effects of heme deficiency on APP processing in the cells overexpressing APP and neuronal cell expressing APP endogenously. It has to be noted that inhibition of heme synthesis at the terminal stage of the pathway (FECH) inevitably creates an accumulation of protoporphyrin precursors, which can be toxic and also can have an effect on APP. Additionally, $10\,\mu M$ is a very high concentration considering that 1-μM NMP IX is enough to reduce the rate of heme synthesis in neurons by ~ 60% (92) and a drastic inhibition may quickly cause mitochondrial failure and trigger secondary changes in various signaling pathways, especially those which are ATP- and cGMP-dependent.

Recent work by Dwyer and colleagues (136) showed that the relative expression of *ALAS1* mRNA was reduced by about 90% in AD brains compared to controls, estimated by semi-quantitative RT-PCR. Similarly, the relative expression of *PBGD* mRNA estimated by quantitative real-time PCR was reduced by about 60%; in contrast, the relative expression of *ALAD* showed no significant changes. The authors suggested that down-regulation of *ALAS1* can be attributed to increased levels of heme-*b* in AD brain. Indeed, previous work by Atamna reported an increase to

250% in heme-b and a 2.9-fold decrease of the heme-a/heme-b ratio in the temporal lobe of AD patients compared with nondemented controls (132). However, down-regulation of heme synthesis (the rate limiting step *ALAS1*) would not support Atamna's hypothesis of functional heme deficiency resulted from heme binding to Aβ. If available heme is depleted by sequestration in Aβ aggregates, which are abundant in AD brains, combined with a profound impairment of mitochondrial function where ALAS1 protein acts, this can eventually lead to failure of compensatory mechanisms. Under such circumstances, mechanisms employed to provide a regulatory feedback from free heme levels to *ALAS1* gene expression might be corrupted.

Heme in circadian pathways

In mammals, the circadian mechanism governing expression of many genes in periphery organs is controlled by a master clock in the suprachiasmatic nucleus of the hypothalamus (137–139). Both *ALAS1* and *ALAS2* expression in mouse regulating heme synthesis, for instance, appear to be governed by such a mechanism apparently via operation of the *PER1* and *PER2* genes. This implies that heme supply and hemoproteins in the brain might be subject to circadian regulation. In turn, heme supply itself might regulate the clock by a feedback mechanism via binding to PAS domain proteins in the case of nPAS2, allowing subsequent binding of CO (also produced from heme by the action of HO), or by regulating the stability of PER2 protein (110, 140–144). This suggests a potential route for pharmacological intervention. Another important component of the clock mechanism is REV-ERBα, also binding heme, a nuclear steroid receptor regulating transcription of the essential clock transcriptional activator, bMAL1(84–85). Thus, the relationship between ALAS1 expression, heme availability and clock proteins controlling the circadian cycle is complex, intriguing and of fundamental importance. However, recent studies have cautioned interpretation of findings on the role of heme in this field with respect to PER and nPAS2 (145). Heme is notoriously sticky and heme bindings to proteins may not always be specific or with the greatest affinity. Even so, the evidence for an important role for heme and its synthesis in circadian rhythm still seems robust including the participation of REV-ERBα.

References

1. Ajioka RS, Phillips JD, Kushner JP. 2006. Biosynthesis of heme in mammals. *Biochim Biophys Acta* 1763: 723–736.

2. Nagaiah K, MacDonnell P, Guroff G. 1977. Induction of tyrosine hydroxlase synthesis in rat superior cervical ganglia *in vitro* by nerve growth factor and dexamethasone. *Biochem Biophys Res Commun* 75: 832–837.

3. Sassa S, Nagai T. 1996. The role of heme in gene expression. *Int J Hematol* 63: 167–178.

4. Badminton MN, Elder GH. 2005. Molecular mechanisms of dominant expression in porphyria. *J Inherit Metab Dis* 28: 277–286.

5. Whatley SD, Ducamp S, Gouya L, Grandchamp B, Beaumont C, Badminton MN, Elder GH, Holme SA, Anstey AV, Parker M, Corrigall AV, Meissner PN, Hift RJ, Marsden JT, Ma Y, Mieli-Vergani G, Deybach JC, Puy H. 2008. C-terminal deletions in the ALAS2 gene lead to gain of function and cause X-linked dominant protoporphyria without anemia or iron overload. *Am J Hum Genet* 83: 408–414.

6. Handschin C, Lin J, Rhee J, Peyer AK, Chin S, Wu PH, Meyer UA, Spiegelman BM. 2005. Nutritional regulation of hepatic heme biosynthesis and porphyria through PGC-1alpha. *Cell* 122: 505–515.

7. Buerk DG, Saidel GM. 1977. A comparison of two nonclassical models for oxygen consumption in brain and liver tissue. *Adv Exp Med Biol* 94: 225–232.

8. Marks GS. 1985. Exposure to toxic agents: The heme biosynthetic pathway and hemoproteins as indicator. *Crit Rev Toxicol* 15: 151–179.

9. Smith AG, Elder GH. 2010. Complex gene-chemical interactions: Hepatic uroporphyria as a paradigm. *Chem Res Toxicol* 23: 712–723.

10. Moore MR, Goldberg A, Yeung-Laiwah AA. 1987. Lead effects on the heme biosynthetic pathway. Relationship to toxicity. *Ann NY Acad Sci* 514: 191–203.

11. Opler MG, Brown AS, Graziano J, Desai M, Zheng W, Schaefer C, Factor-Litvak P, Susser ES. 2004. Prenatal lead exposure, delta-aminolevulinic acid, and schizophrenia. *Environ Health Perspect* 112: 548–552.

12. De Matteis F, Zetterlund P, Wetterberg L. 1981. Brain 5-aminolaevulinate synthase. Developmental aspects and evidence for regulatory role. *Biochem J* 196: 811–817.

13. Tschudy DP, Hess RA, Frykholm BC. 1981. Inhibition of delta-aminolevulinic acid dehydrase by 4,6-dioxoheptanoic acid. *J Biol Chem* 256: 9915–9923.

14. Kang K, Anderson-Burham LA, Bloomer JR. 1987. Effect of succinylacetone administration on brain heme metabolism and behavior in mice. *Biochem Pharmacol* 36: 3084–3086.

15. Paterniti JR, Jr., Lin CI, Beattie DS. 1978. Delta-aminolevulinic acid synthetase: Regulation of activity in various tissues of the aging rat. *Arch Biochem Biophys* 191: 792–797.

16. Litman DA, Correia MA. 1983. L-tryptophan: A common denominator of biochemical and neurological events of acute hepatic porphyria? *Science* 222: 1031–1033.

17. De Matteis F, Ray DE. 1982. Studied on cerebellar haem metabolism in the rat *in vivo*. *J Neurochem* 39: 551–556.

18. Ingi T, Chiang G, Ronnett GV. 1996. The regulation of heme turnover and carbon monoxide biosynthesis in cultured primary rat olfactory receptor neurons. *J Neurosci* 16: 5621–5628.

19. Kumar S, Bandyopadhyay U. 2005. Free heme toxicity and its detoxification systems in human. *Toxicol Lett* 157: 175–188.

20. Ribo M, Grotta JC. 2006. Latest advances in intracerebral hemorrhage. *Curr Neurol Neurosci Rep* 6: 17–22.

21. Wang J, Dore S. 2007. Inflammation after intracerebral hemorrhage. *J Cereb Blood Flow Metab* 27: 894–908.

22. Kaptain GJ, Lanzino G, Kassell NF. 2000. Subarachnoid haemorrhage: Epidemiology, risk factors, and treatment options. *Drugs Aging* 17: 183–199.

23. Hua Y, Xi G, Keep RF, Hoff JT. 2000. Complement activation in the brain after experimental intracerebral hemorrhage. *J Neurosurg* 92: 1016–1022.

24. Wagner KR, Sharp FR, Ardizzone TD, Lu A, Clark JF. 2003. Heme and iron metabolism: Role in cerebral hemorrhage. *J Cereb Blood Flow Metab* 23: 629–652.

25. Umbreit J. 2007. Methemoglobin — it's not just blue: A concise review. *Am J Hematol* 82: 134–144.

26. Pohlman TH, Harlan JM. 2000. Adaptive responses of the endothelium to stress. *J Surg Res* 89: 85–119.

27. Arruda MA, Rossi AG, de Freitas MS, Barja-Fidalgo C, Graca-Souza AV. 2004. Heme inhibits human neutrophil apoptosis: Involvement of phosphoinositide 3-kinase, MAPK, and NF-kappaB. *J Immunol* 173: 2023–2030.

28. Goldstein L, Teng ZP, Zeserson E, Patel M, Regan RF. 2003. Hemin induces an iron-dependent, oxidative injury to human neuron-like cells. *J Neurosci Res* 73: 113–121.

29. Qureshi AI, Tuhrim S, Broderick JP, Batjer HH, Hondo H, Hanley DF. 2001. Spontaneous intracerebral hemorrhage. *N Engl J Med* 344: 1450–1460.

30. Li RC, Saleem S, Zhen G, Cao W, Zhuang H, Lee J, Smith A, Altruda F, Tolosano E, Dore S. 2009. Heme-hemopexin complex attenuates neuronal cell death and stroke damage. *J Cereb Blood Flow Metab* 29: 953–964.

31. Turner CP, Bergeron M, Matz P, Zegna A, Noble LJ, Panter SS, Sharp FR. 1998. Heme oxygenase-1 is induced in glia throughout brain by subarachnoid hemoglobin. *J Cereb Blood Flow Metab* 18: 257–273.

32. Matz PG, Weinstein PR, Sharp FR. 1997. Heme oxygenase-1 and heat shock protein 70 induction in glia and neurons throughout rat brain after experimental intracerebral hemorrhage. *Neurosurgery* 40: 152–160; discussion 60–62.

33. Aft RL, Mueller GC. 1983. Hemin-mediated DNA strand scission. *J Biol Chem* 258: 12069–12072.

34. Sassa S. 2004. Why heme needs to be degraded to iron, biliverdin IXalpha, and carbon monoxide? *Antioxid Redox Signal* 6: 819–824.

35. Clark JF, Loftspring M, Wurster WL, Beiler S, Beiler C, Wagner KR, Pyne-Geithman GJ. 2008. Bilirubin oxidation products, oxidative stress, and intracerebral hemorrhage. *Acta Neurochir Suppl* 105: 7–12.

36. Miller YI, Felikman Y, Shaklai N. 1996. Hemoglobin induced apolipoprotein B crosslinking in low-density lipoprotein peroxidation. *Arch Biochem Biophys* 326: 252–260.

37. Valko M, Morris H, Cronin MT. 2005. Metals, toxicity and oxidative stress. *Curr Med Chem* 12: 1161–1208.

38. Regan RF, Panter SS. 1996. Hemoglobin potentiates excitotoxic injury in cortical cell culture. *J Neurotrauma* 13: 223–231.

39. Xi G, Keep RF, Hoff JT. 1998. Erythrocytes and delayed brain edema formation following intracerebral hemorrhage in rats. *J Neurosurg* 89: 991–996.

40. Hua Y, Nakamura T, Keep RF, Wu J, Schallert T, Hoff JT, Xi G. 2006. Long-term effects of experimental intracerebral hemorrhage: The role of iron. *J Neurosurg* 104: 305–312.

41. Regan RF, Panter SS. 1993. Neurotoxicity of hemoglobin in cortical cell culture. *Neurosci Lett* 153: 219–222.

42. Letarte PB, Lieberman K, Nagatani K, Haworth RA, Odell GB, Duff TA. 1993. Hemin: Levels in experimental subarachnoid hematoma and effects on dissociated vascular smooth-muscle cells. *J Neurosurg* 79: 252–255.

43. Jaremko KM, Chen-Roetling J, Chen L, Regan RF. 2010. Accelerated hemolysis and neurotoxicity in neuron-glia-blood clot co-cultures. *J Neurochem* 114: 1063–1073.

44. Vecchione C, Frati A, Di Pardo A, Cifelli G, Carnevale D, Gentile MT, Carangi R, Landolfi A, Carullo P, Bettarini U, Antenucci G, Mascio G,

Busceti CL, Notte A, Maffei A, Cantore GP, Lembo G. 2009. Tumor necrosis factor-alpha mediates hemolysis-induced vasoconstriction and the cerebral vasospasm evoked by subarachnoid hemorrhage. *Hypertension* 54: 150–156.

45. Sadrzadeh SM, Eaton JW. 1988. Hemoglobin-mediated oxidant damage to the central nervous system requires endogenous ascorbate. *J Clin Invest* 82: 1510–1515.

46. Haorah J, Ramirez SH, Schall K, Smith D, Pandya R, Persidsky Y. 2007. Oxidative stress activates protein tyrosine kinase and matrix metalloproteinases leading to blood-brain barrier dysfunction. *J Neurochem* 101: 566–576.

47. Wang J, Tsirka SE. 2005. Neuroprotection by inhibition of matrix metalloproteinases in a mouse model of intracerebral haemorrhage. *Brain* 128: 1622–1633.

48. Katsu M, Niizuma K, Yoshioka H, Okami N, Sakata H, Chan PH. 2010. Hemoglobin-induced oxidative stress contributes to matrix metalloproteinase activation and blood-brain barrier dysfunction *in vivo*. *J Cereb Blood Flow Metab* 30: 1939–1950.

49. Xue M, Fan Y, Liu S, Zygun DA, Demchuk A, Yong VW. 2009. Contributions of multiple proteases to neurotoxicity in a mouse model of intracerebral haemorrhage. *Brain* 132: 26–36.

50. Wang X, Mori T, Sumii T, Lo EH. 2002. Hemoglobin-induced cytotoxicity in rat cerebral cortical neurons: Caspase activation and oxidative stress. *Stroke* 33: 1882–1888.

51. Burmester T, Weich B, Reinhardt S, Hankeln T. 2000. A vertebrate globin expressed in the brain. *Nature* 407: 520–523.

52. Herold S, Fago A, Weber RE, Dewilde S, Moens L. 2004. Reactivity studies of the Fe(III) and Fe(II)NO forms of human neuroglobin reveal a potential role against oxidative stress. *J Biol Chem* 279: 22841–22847.

53. Ogawa K, Sun J, Taketani S, Nakajima O, Nishitani C, Sassa S, Hayashi N, Yamamoto M, Shibahara S, Fujita H, Igarashi K. 2001. Heme mediates derepression of Maf recognition element through direct binding to transcription repressor Bach1. *EMBO J* 20: 2835–2843

54. Fago A, Hundahl C, Dewilde S, Gilany K, Moens L, Weber RE. 2004. Allosteric regulation and temperature dependence of oxygen binding in human neuroglobin and cytoglobin. Molecular mechanisms and physiological significance. *J Biol Chem* 279: 44417–44426

55. Bonding SH, Henty K, Dingley AJ, Brittain T. 2008. The binding of cytochrome c to neuroglobin: A docking and surface plasmon resonance study. *Int J Biol Macromol* 43: 295–299.

56. Fago A, Mathews AJ, Brittain T. 2008. A role for neuroglobin: Resetting the trigger level for apoptosis in neuronal and retinal cells. *IUBMB Life* 60: 398–401.

57. Sun Y, Jin K, Peel A, Mao XO, Xie L, Greenberg DA. 2003. Neuroglobin protects the brain from experimental stroke *in vivo*. *Proc Natl Acad Sci U S A* 100: 3497–3500.

58. Li RC, Morris MW, Lee SK, Pouranfar F, Wang Y, Gozal D. 2008. Neuroglobin protects PC12 cells against oxidative stress. *Brain Res* 1190: 159–166.

59. Reuben RC, Khanna PL, Gazitt Y, Breslow R, Rifkind RA, Marks PA. 1978. Inducers of erythroleukemic differentiation. Relationship of structure to activity among planar-polar compounds. *J Biol Chem* 253: 4214–4218.

60. Li RC, Guo SZ, Lee SK, Gozal D. 2010. Neuroglobin protects neurons against oxidative stress in global ischemia. *J Cereb Blood Flow Metab* 30: 1874–1882.

61. Wakasugi K, Kitatsuji C, Morishima I. 2005. Possible neuroprotective mechanism of human neuroglobin. *Ann N Y Acad Sci* 1053: 220–230.

62. Schelshorn DW, Schneider A, Kuschinsky W, Weber D, Kruger C, Dittgen T, Burgers HF, Sabouri F, Gassler N, Bach A, Maurer MH. 2009. Expression of hemoglobin in rodent neurons. *J Cereb Blood Flow Metab* 29: 585–595.

63. Biagioli M, Pinto M, Cesselli D, Zaninello M, Lazarevic D, Roncaglia P, Simone R, Vlachouli C, Plessy C, Bertin N, Beltrami A, Kobayashi K, Gallo V, Santoro C, Ferrer I, Rivella S, Beltrami CA, Carninci P, Raviola E, Gustincich S. 2009. Unexpected expression of alpha- and beta-globin in mesencephalic dopaminergic neurons and glial cells. *Proc Natl Acad Sci U S A* 106: 15454–15459.

64. Nakamura A, Fujiwara S, Ishijima T, Okada S, Nakai Y, Matsumoto I, Misaka T, Abe K. 2010. Neuron differentiation-related genes are up-regulated in the hypothalamus of odorant-inhaling rats subjected to acute restraint stress. *J Agric Food Chem* 58: 7922–7929.

65. McCoubrey WK, Jr., Huang TJ, Maines MD. 1997. Isolation and characterization of a cDNA from the rat brain that encodes hemoprotein heme oxygenase-3. *Eur J Biochem* 247: 725–732.

66. Hayashi S, Omata Y, Sakamoto H, Higashimoto Y, Hara T, Sagara Y, Noguchi M. 2004. Characterization of rat heme oxygenase-3 gene. Implication of processed pseudogenes derived from heme oxygenase-2 gene. *Gene* 336: 241–250.

67. Verma A, Hirsch DJ, Glatt CE, Ronnett GV, Snyder SH. 1993. Carbon monoxide: A putative neural messenger. *Science* 259: 381–384.

68. Maines MD. 2000. The heme oxygenase system and its functions in the brain. *Cell Mol Biol (Noisy-le-grand)* 46: 573–585.

69. Schipper HM, Song W, Zukor H, Hascalovici JR, Zeligman D. 2009. Heme oxygenase-1 and neurodegeneration: Expanding frontiers of engagement. *J Neurochem* 110: 469–485.

70. Takeda A, Perry G, Abraham NG, Dwyer BE, Kutty RK, Laitinen JT, Petersen RB, Smith MA. 2000. Overexpression of heme oxygenase in neuronal cells, the possible interaction with Tau. *J Biol Chem* 275: 5395–5399.

71. Cuadrado A, Rojo AI. 2008. Heme oxygenase-1 as a therapeutic target in neurodegenerative diseases and brain infections. *Curr Pharm Des* 14: 429–442.

72. Scapagnini G, Colombrita C, Amadio M, D'Agata V, Arcelli E, Sapienza M, Quattrone A, Calabrese V. 2006. Curcumin activates defensive genes and protects neurons against oxidative stress. *Antioxid Redox Signal* 8: 395–403.

73. Sakoda E, Igarashi K, Sun J, Kurisu K, Tashiro S. 2008. Regulation of heme oxygenase-1 by transcription factor Bach1 in the mouse brain. *Neurosci Lett* 440: 160–165.

74. Kannan M, Steinert JR, Forsythe ID, Smith AG, Chernova T. 2010. Mevastatin accelerates loss of synaptic proteins and neurite degeneration in aging cortical neurons in a heme-independent manner. *Neurobiol Aging* 31: 1543–1553.

75. Schipper HM, Gupta A, Szarek WA. 2009. Suppression of glial HO-1 activity as a potential neurotherapeutic intervention in AD. *Curr Alzheimer Res* 6: 424–430.

76. Dore S, Takahashi M, Ferris CD, Zakhary R, Hester LD, Guastella D, Snyder SH. 1999. Bilirubin, formed by activation of heme oxygenase-2, protects neurons against oxidative stress injury. *Proc Natl Acad Sci U S A* 96: 2445–2450.

77. Regan RF, Kumar N, Gao F, Guo Y. 2002. Ferritin induction protects cortical astrocytes from heme-mediated oxidative injury. *Neuroscience* 113: 985–994.

78. Rogers B, Yakopson V, Teng ZP, Guo Y, Regan RF. 2003. Heme oxygenase-2 knockout neurons are less vulnerable to hemoglobin toxicity. *Free Radic Biol Med* 35: 872–881.

79. Chang EF, Wong RJ, Vreman HJ, Igarashi T, Galo E, Sharp FR, Stevenson DK, Noble-Haeusslein LJ. 2003. Heme oxygenase-2 protects against lipid peroxidation-mediated cell loss and impaired motor recovery after traumatic brain injury. *J Neurosci* 23: 3689–3696.

80. Ingi T, Cheng J, Ronnett GV. 1996. Carbon monoxide: An endogenous modulator of the nitric oxide-cyclic GMP signaling system. *Neuron* 16: 835–842.

81. Chen JJ, London IM. 1995. Regulation of protein synthesis by heme-regulated eIF-2 alpha kinase. *Trends Biochem Sci* 20: 105–108.

82. Wang S, Publicover S, Gu Y. 2009. An oxygen-sensitive mechanism in regulation of epithelial sodium channel. *Proc Natl Acad Sci U S A* 106: 2957–2962.

83. Tang XD, Xu R, Reynolds MF, Garcia ML, Heinemann SH, Hoshi T. 2003. Haem can bind to and inhibit mammalian calcium-dependent Slo1 BK channels. *Nature* 425: 531–535.

84. Yin L, Wu N, Curtin JC, Qatanani M, Szwergold NR, Reid RA, Waitt GM, Parks DJ, Pearce KH, Wisely GB, Lazar MA. 2007. Rev-erbalpha, a heme sensor that coordinates metabolic and circadian pathways. *Science* 318: 1786–1789.

85. Raghuram S, Stayrook KR, Huang P, Rogers PM, Nosie AK, McClure DB, Burris LL, Khorasanizadeh S, Burris TP, Rastinejad F. 2007. Identification of heme as the ligand for the orphan nuclear receptors REV-ERBalpha and REV-ERBbeta. *Nat Struct Mol Biol* 14: 1207–1213.

86. Faller M, Matsunaga M, Yin S, Loo JA, Guo F. 2007. Heme is involved in microRNA processing. *Nat Struct Mol Biol* 14: 23–29.

87. Roberts AG, Elder GH. 2001. Alternative splicing and tissue-specific transcription of human and rodent ubiquitous 5-aminolevulinate synthase (ALAS1) genes. *Biochim Biophys Acta* 1518: 95–105.

88. Meyer RP, Gehlhaus M. 2010. A role for CYP in the drug-hormone crosstalk of the brain. *Expert Opin Drug Metab Toxicol* 6: 675–687.

89. Hamilton JW, Bement WJ, Sinclair PR, Sinclair JF, Alcedo JA, Wetterhahn KE. 1991. Heme regulates hepatic 5-aminolevulinate synthase mRNA expression by decreasing mRNA half-life and not by altering its rate of transcription. *Arch Biochem Biophys* 289: 387–392.

90. Hayashi N, Terasawa M, Yamauchi K, Kikuchi G. 1980. Effects of hemin on the synthesis and intracellular translocation of delta-aminolevulinate synthase in the liver of rats treated with 3,5-dicarbethoxy-1,4-dihydrocollidine. *J Biochem* 88: 1537–1543.

91. Atamna H, Liu J, Ames BN. 2001. Heme deficiency selectively interrupts assembly of mitochondrial complex IV in human fibroblasts: Revelance to aging. *J Biol Chem* 276: 48410–48416.

92. Chernova T, Nicotera P, Smith AG. 2006. Heme deficiency is associated with senescence and causes suppression of N-methyl-D-aspartate receptor

subunits expression in primary cortical neurons. *Mol Pharmacol* 69: 697–705.

93. Chernova T, Steinert JR, Guerin CJ, Nicotera P, Forsythe ID, Smith AG. 2007. Neurite degeneration induced by heme deficiency mediated via inhibition of NMDA receptor-dependent extracellular signal-regulated kinase 1/2 activation. *J Neurosci* 27: 8475–8485.

94. Sengupta A, Hon T, Zhang L. 2005. Heme deficiency suppresses the expression of key neuronal genes and causes neuronal cell death. *Brain Res Mol Brain Res* 137: 23–30.

95. Davies R, Schuurman A, Barker CR, Clothier B, Chernova T, Higginson FM, Judah DJ, Dinsdale D, Edwards RE, Greaves P, Gant TW, Smith AG. 2005. Hepatic gene expression in protoporphyic Fech mice is associated with cholestatic injury but not a marked depletion of the heme regulatory pool. *Am J Pathol* 166: 1041–1053.

96. Tutois S, Montagutelli X, Da Silva V, Jouault H, Rouyer-Fessard P, Leroy-Viard K, Guenet JL, Nordmann Y, Beuzard Y, Deybach JC. 1991. Erythropoietic protoporphyria in the house mouse. A recessive inherited ferrochelatase deficiency with anemia, photosensitivity, and liver disease. *J Clin Invest* 88: 1730–1736.

97. Wilks A, Ortiz de Montellano PR, Sun J, Loehr TM. 1996. Heme oxygenase (HO-1): His-132 stabilizes a distal water ligand and assists catalysis. *Biochemistry* 35: 930–936.

98. Zhu Y HT, Ye W, Zhang L. 2002. Heme deficiency interferes with the Ras-mitogen-activated protein kinase signaling pathway and expression of a subset of neuronal genes. *Cell Growth Differ.* 13: 431–439.

99. Hardingham GE. 2006. Pro-survival signalling from the NMDA receptor. *Biochem Soc Trans* 34: 936–938.

100. Ishii DN, Maniatis GM. 1978. Haemin promotes rapid neurite outgrowth in cultured mouse neuroblastoma cells. *Nature* 274: 372–374.

101. Greengard P. 1978. Phosphorylated proteins as physiological effectors. *Science* 199: 146–152.

102. Shinjyo N, Kita K. 2006. Up-regulation of heme biosynthesis during differentiation of Neuro2a cells. *J Biochem* 139: 373–381.

103. Shinjyo N, Kita K. 2007. Relationship between reactive oxygen species and heme metabolism during the differentiation of Neuro2a cells. *Biochem Biophys Res Commun* 358: 130–135.

104. Atamna H, Killilea DW, Killilea AN, Ames BN. 2002. Heme deficiency may be a factor in the mitochondrial and neuronal decay of aging. *Proc Natl Acad Sci U S A* 99: 14807–14812.

105. Kimura I, Yoshioka M, Konishi M, Miyake A, Itoh N. 2005. Neudesin, a novel secreted protein with a unique primary structure and neurotrophic activity. *J Neurosci Res* 79: 287–294.

106. Kimura I, Nakayama Y, Yamauchi H, Konishi M, Miyake A, Mori M, Ohta M, Itoh N, Fujimoto M. 2008. Neurotrophic activity of neudesin, a novel extracellular heme-binding protein, is dependent on the binding of heme to its cytochrome b5-like heme/steroid-binding domain. *J Biol Chem* 283: 4323–4331.

107. Kimura I, Nakayama Y, Konishi M, Kobayashi T, Mori M, Ito M, Hirasawa A, Tsujimoto G, Ohta M, Itoh N, Fujimoto M. 2010. Neuferricin, a novel extracellular heme-binding protein, promotes neurogenesis. *J Neurochem* 112: 1156–1167.

108. Rodgers KR. 1999. Heme-based sensors in biological systems. *Curr Opin Chem Biol* 3: 158–167.

109. Gilles-Gonzalez MA, Gonzalez G. 2005. Heme-based sensors: Defining characteristics, recent developments, and regulatory hypotheses. *J Inorg Biochem* 99: 1–22.

110. Dioum EM, Rutter J, Tuckerman JR, Gonzalez G, Gilles-Gonzalez MA, McKnight SL. 2002. NPAS2: A gas-responsive transcription factor. *Science* 298: 2385–2387.

111. Boehning D, Snyder SH. 2002. Circadian rhythms. Carbon monoxide and clocks. *Science* 298: 2339–2340.

112. Boehning D, Snyder SH. 2003. Novel neural modulators. *Annu Rev Neurosci* 26: 105–131.

113. Bitar MS, Shapiro BH. 1987. Aberration of heme and hemoprotein in aged female rats. *Mech Ageing Dev* 38: 189–197.

114. Meyer RP, Podvinec M, Meyer UA. 2002. Cytochrome P450 CYP1A1 accumulates in the cytosol of kidney and brain and is activated by heme. *Mol Pharmacol* 62: 1061–1067.

115. Correia MA, Meyer UA. 1975. Apocytochrome P-450: Reconstitution of functional cytochrome with hemin *in vitro*. *Proc Natl Acad Sci U S A* 72: 400–404.

116. Podvinec M, Handschin C, Looser R, Meyer UA. 2004. Identification of the xenosensors regulating human 5-aminolevulinate synthase. *Proc Natl Acad Sci U S A* 101: 9127–9132.

117. Fraser DJ, Zumsteg A, Meyer UA. 2003. Nuclear receptors constitutive androstane receptor and pregnane X receptor activate a drug-responsive enhancer of the murine 5-aminolevulinic acid synthase gene. *J Biol Chem* 278: 39392–39401.

118. Lin JH, Lu AY. 2001. Interindividual variability in inhibition and induction of cytochrome P450 enzymes. *Annu Rev Pharmacol Toxicol* 41: 535–567.

119. Miksys S, Tyndale RF. 2004. The unique regulation of brain cytochrome P450 2 (CYP2) family enzymes by drugs and genetics. *Drug Metab Rev* 36: 313–333.

120. Siegle I, Fritz P, Eckhardt K, Zanger UM, Eichelbaum M. 2001. Cellular localization and regional distribution of CYP2D6 mRNA and protein expression in human brain. *Pharmacogenetics* 11: 237–245.

121. Gervasini G, Carrillo JA, Benitez J. 2004. Potential role of cerebral cytochrome P450 in clinical pharmacokinetics: Modulation by endogenous compounds. *Clin Pharmacokinet* 43: 693–706.

122. Kennerfalk A, Ruigomez A, Wallander MA, Wilhelmsen L, Johansson S. 2002. Geriatric drug therapy and healthcare utilization in the United kingdom. *Ann Pharmacother* 36: 797–803.

123. Meyer RP, Lindberg RL, Hoffmann F, Meyer UA. 2005. Cytosolic persistence of mouse brain CYP1A1 in chronic heme deficiency. *Biol Chem* 386: 1157–1164.

124. Yau JL, Noble J, Graham M, Seckl JR. 2006. Central administration of a cytochrome P450-7B product 7 alpha-hydroxypregnenolone improves spatial memory retention in cognitively impaired aged rats. *J Neurosci* 26: 11034–11040.

125. Rose KA, Stapleton G, Dott K, Kieny MP, Best R, Schwarz M, Russell DW, Bjorkhem I, Seckl J, Lathe R. 1997. Cyp7b, a novel brain cytochrome P450, catalyzes the synthesis of neurosteroids 7alpha-hydroxy dehydroepiandrosterone and 7alpha-hydroxy pregnenolone. *Proc Natl Acad Sci U S A* 94: 4925–4930.

126. Magnusson KR, Nelson SE, Young AB. 2002. Age-related changes in the protein expression of subunits of the NMDA receptor. *Brain Res Mol Brain Res* 99: 40–45.

127. Raff MC, Whitmore AV, Finn JT. 2002. Axonal self-destruction and neurodegeneration. *Science* 296: 868–871.

128. Subbarao KV, Richardson JS. 1990. Iron-dependent peroxidation of rat brain: a regional study. *J Neurosci Res* 26: 224–232.

129. Pamplona R, Dalfo E, Ayala V, Bellmunt MJ, Prat J, Ferrer I, Portero-Otin M. 2005. Proteins in human brain cortex are modified by oxidation, glycoxidation, and lipoxidation. Effects of Alzheimer disease and identification of lipoxidation targets. *J Biol Chem* 280: 21522–21530.

130. Altamura S, Muckenthaler MU. 2009. Iron toxicity in diseases of aging: Alzheimer's disease, Parkinson's disease and atherosclerosis. *J Alzheimers Dis* 16: 879–895.

131. Atamna H, Frey WH, 2nd, Ko N. 2009. Human and rodent amyloid-beta peptides differentially bind heme: Relevance to the human susceptibility to Alzheimer's disease. *Arch Biochem Biophys* 487: 59–65.

132. Atamna H, Frey WH, 2nd. 2004. A role for heme in Alzheimer's disease: heme binds amyloid beta and has altered metabolism. *Proc Natl Acad Sci U S A* 101: 11153–11158.

133. Scheuermann S, Hambsch B, Hesse L, Stumm J, Schmidt C, Beher D, Bayer TA, Beyreuther K, Multhaup G. 2001. Homodimerization of amyloid precursor protein and its implication in the amyloidogenic pathway of Alzheimer's disease. *J Biol Chem* 276: 33923–33929.

134. Howlett D, Cutler P, Heales S, Camilleri P. 1997. Hemin and related porphyrins inhibit beta-amyloid aggregation. *FEBS Lett* 417: 249–251.

135. Gatta LB, Vitali M, Verardi R, Arosio P, Finazzi D. 2009. Inhibition of heme synthesis alters Amyloid Precursor Protein processing. *J Neural Transm* 116: 79–88.

136. Dwyer BE, Smith MA, Richardson SL, Perry G, Zhu X. 2009. Down-regulation of aminolevulinate synthase, the rate-limiting enzyme for heme biosynthesis in Alzheimer's disease. *Neurosci Lett* 460: 180–184.

137. Zheng B, Albrecht U, Kaasik K, Sage M, Lu W, Vaishnav S, Li Q, Sun ZS, Eichele G, Bradley A, Lee CC. 2001. Nonredundant roles of the mPer1 and mPer2 genes in the mammalian circadian clock. *Cell* 105: 683–694.

138. Panda S, Antoch MP, Miller BH, Su AI, Schook AB, Straume M, Schultz PG, Kay SA, Takahashi JS, Hogenesch JB. 2002. Coordinated transcription of key pathways in the mouse by the circadian clock. *Cell* 109: 307–320.

139. Akhtar RA, Reddy AB, Maywood ES, Clayton JD, King VM, Smith AG, Gant TW, Hastings MH, Kyriacou CP. 2002. Circadian cycling of the mouse liver transcriptome, as revealed by cDNA microarray, is driven by the suprachiasmatic nucleus. *Curr Biol* 12: 540–550.

140. Kaasik K, Lee CC. 2004. Reciprocal regulation of haem biosynthesis and the circadian clock in mammals. *Nature* 430: 467–471.

141. Koudo R, Kurokawa H, Sato E, Igarashi J, Uchida T, Sagami I, Kitagawa T, Shimizu T. 2005. Spectroscopic characterization of the isolated heme-bound PAS-B domain of neuronal PAS domain protein 2 associated with circadian rhythms. *Febs J* 272: 4153–4162.

142. Mukaiyama Y, Uchida T, Sato E, Sasaki A, Sato Y, Igarashi J, Kurokawa H, Sagami I, Kitagawa T, Shimizu T. 2006. Spectroscopic and DNA-binding characterization of the isolated heme-bound basic helix-loop-helix-PAS-A domain of neuronal PAS protein 2 (NPAS2), a transcription activator protein associated with circadian rhythms. *Febs J* 273: 2528–2539.

143. Kitanishi K, Igarashi J, Hayasaka K, Hikage N, Saiful I, Yamauchi S, Uchida T, Ishimori K, Shimizu T. 2008. Heme-binding characteristics of the isolated PAS-A domain of mouse Per2, a transcriptional regulatory factor associated with circadian rhythms. *Biochemistry* 47: 6157–6168.

144. Yang J, Kim KD, Lucas A, Drahos KE, Santos CS, Mury SP, Capelluto DG, Finkielstein CV. 2008. A novel heme-regulatory motif mediates heme-dependent degradation of the circadian factor period 2. *Mol Cell Biol* 28: 4697–4711.

145. Airola MV, Du J, Dawson JH, Crane BR. 2010. Heme binding to the mammalian circadian clock protein period 2 is nonspecific. *Biochemistry* 49: 4327–4338.

Chapter 6

HEME AND microRNA BIOGENESIS

Feng Guo

Overview

microRNAs (miRNAs) are a class of non-protein-coding small RNAs that function in development and cell physiology through regulation of gene expression (1). In this chapter, I describe the discovery of an essential miRNA processing factor, DiGeorge Critical Region 8 (DGCR8), as a previously unknown heme-binding protein. Our biochemical and structural characterization of the heme-DGCR8 interaction suggests that heme is involved in the regulation of miRNA processing. Future directions of our investigation are discussed.

miRNAs and the miRNA biogenesis pathway

miRNAs are ~22-nucleotide single-stranded RNAs containing a 5′-phosphate. miRNAs are found in a wide range of organisms, including mammals, amphibians, fish, insects, worms and plants. Over 700 miRNAs have been identified in humans. They regulate the expression of a large number of protein-coding genes through translational repression or degradation of their messenger RNAs (mRNAs) (2, 3). miRNAs and transcription factors constitute gene regulatory networks essential for developmental programming and physiological responses. For example, miR-196 negatively regulates the expression of the transcription factor Bach1 (4). Bach1 is directly regulated by heme and is essential for heme-dependent regulation of the cytoprotective heme-degrading enzyme heme oxygenase 1 (HMOX1) during stress responses and of β-globin during erythroid differentiation (5). miRNAs are bound to the Argounate proteins in an effector complex called miRNA-induced silencing complex (miRISC) and specifically recognize target mRNAs through pairing interactions. Plant miRNAs are nearly completely complementary to target mRNAs. In contrast, animal miRNAs most often form partial pairing interactions with target mRNAs, with the nucleotides at positions 2–8 (counting from its 5′-end, called the seed region) being most important for target recognition.

Most miRNAs are transcribed by RNA polymerase II as primary transcripts (pri-miRNAs). Pri-miRNAs contain characteristic 30-base-pair hairpin secondary structures in which mature miRNAs reside. A pri-miRNA can be an independent transcript, or a pre-mRNA with miRNAs located in its introns or, in some cases, exons. Pri-miRNAs are processed through sequential cleavages (Fig. 1) (6, 7). First, in the nucleus, a pri-mRNA is processed into a 65–70 nucleotide RNA intermediate called a pre-miRNA by a complex called Microprocessor. The Microprocessor minimally contains Drosha (an RNase III-type endonuclease) (8) and DGCR8 (its insect and worm homologs are called Pasha and Pash-1, respectively) (9–12). Drosha cleaves a pri-miRNA at two sites ~ 10 base pairs from the bottom of the hairpin, with the central product being the pre-miRNA. After the pre-miRNAs are exported to the cytosol, they are cleaved further by the Dicer ribonuclease. The resulting duplex miRNAs are unwound and incorporated into the miRISC complexes (13, 14).

miRNAs have been shown to play important roles in diseases such as cancer (15). In addition to altered expression of the individual miRNAs, global decreases of miRNA levels have been demonstrated in normal development and in many cancers (16, 17). Knockdown of *DGCR8* using RNA interference enhanced cellular transformation and tumorigenesis (18),

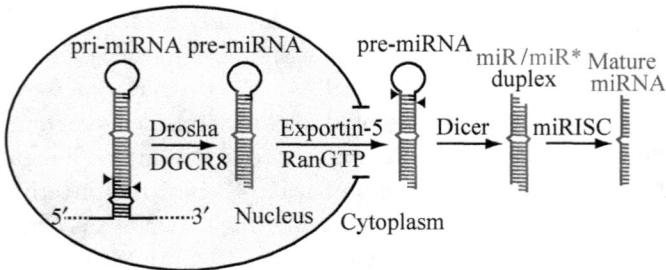

Fig. 1. miRNA biogenesis pathway in animals.

demonstrating a contribution of the global down-regulation of miRNAs in cancer development. However, the mechanism for the global suppression of miRNA expression is not well understood.

DiGeorge Critical Region 8

DGCR8 is essential for the processing of pri-miRNAs *in vitro* and *in vivo* (9, 10, 19, 20). The *DGCR8* gene, along with ~30 other genes, is located in chromosome region 22q11.2 that is heterozygously deleted in the DiGeorge Syndrome patients (21). A mouse model showed that haploin-sufficiency of the Dgcr8 gene results in abnormal miRNA biogenesis in the brain and contributes to the behavioral and neuronal deficits associated with the DiGeorge Syndrome (20). Thus, the expression level of DGCR8 appears to be critical for proper processing of pri-miRNAs. Indeed, the expression and activity of DGCR8 are regulated through multiple mechanisms. For example, Drs. Kim and Gregory's laboratories reported that the mRNA of DGCR8 contains two miRNA-like hairpins that are bound and cleaved by DGCR8 and Drosha, providing a feedback regulation for the expression of DGCR8 (22, 23).

The 773-residue DGCR8 protein contains a WW motif in its central region and two double-stranded RNA-binding domains (dsRBDs) in its C-terminal region (Fig. 2). It binds to pri-miRNAs even in the absence of Drosha (24–26). Its C-terminal tail (CTT) is required for association with Drosha and for assembly of the proper higher-order structure of DGCR8 upon binding pri-miRNAs ((27) and Faller *et al.* submitted). A major function of DGCR8 is to help Drosha recognize pri-miRNA substrates for processing.

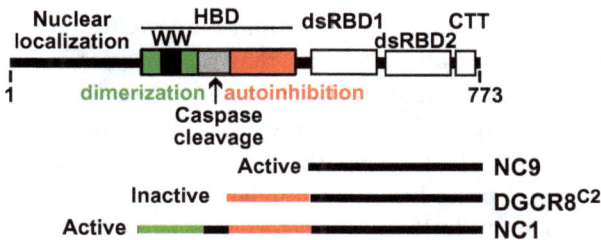

Fig. 2. Domain structure of DGCR8.

DGCR8 binds heme

We initiated the DGCR8 project aiming to answer the question: how are the several hundred pri-miRNAs in humans specifically recognized by their processing factors? Although all the miRNA hairpins contain ~30-bp double-stranded regions, they contain structural irregularities such as bulges and internal loops. These irregularities differ in terms of the types and locations in the hairpins. The goal of the project was to obtain three-dimensional structures of tertiary complexes containing Drosha, DGCR8 and pri-miRNAs, with the hope to reveal the "three-dimensional codes" for recognition of pri-miRNAs. My postdoctoral colleague, Dr. Michael Faller, expressed a truncated active DGCR8, called NC1 and containing residues 276–751, in *E. coli*. He found that NC1 directly binds pri-miRNAs with high cooperativity and assembles into a higher-order complex with pri-miRNAs (25). In his attempts to crystallize the DGCR8-RNA complex, he noticed that the highly purified NC1 protein had a yellow color. He followed his curiosity about this color and eventually identified the origin of the color as a heme cofactor bound to the NC1 protein.

The identity of heme as the cofactor bound to recombinant DGCR8 has been confirmed using multiple methods:

(1) Fluorescent spectrometry. Both heme and flavin derivatives, such as flavin adenine dinucleotide (FAD), can absorb light at a peak wavelength of 450 nm. However, they differ in that flavin has strong fluorescent signals whereas heme does not fluoresce. Both excitation and emission fluorescence spectra of heme-bound DGCR8 showed little fluorescent signals, ruling out the possibility of the cofactor bound to DGCR8 being a flavin.

(2) HPLC. The cofactor was extracted from DGCR8 and was characterized using HPLC. This experiment identified the cofactor as heme in several ways: the cofactor elutes from the reverse-phase column at the same volume as purified hemin; the cofactor and the DGCR8 protein elute separately from the column, suggesting that they are not covalently linked; upon dissociation from the protein, the cofactor loses the absorption peak at 450 nm, instead obtaining a strong absorption at 398 nm, similar to hemin. By integrating the peak volumes in the HPLC chromatograms and using hemin as standards, the extinction coefficient of the heme bound to DGCR8 was determined to be $58\,000\ \text{M}^{-1}\,\text{cm}^{-1}$.

(3) Mass spectrometry. A peak with a mass/charge ratio (m/z) of 616, equal to the molecular weight of a free heme molecule, was observed

in electrospray ionization (ESI) mass spectrometry, performed in denaturing conditions. In native conditions, ESI mass spectra revealed peaks corresponding to a complex with two NC1 subunits and one heme cofactor.

(4) Bacterial expression with and without a heme biosynthesis intermediate, δ-aminolevulinic acid (δ-ALA). In the absence of δ-ALA, overexpression of DGCR8 in *E. coli* produces both heme-bound and heme-free forms of the protein. When 1 mM of δ-ALA is added at the time of induction, only heme-bound DGCR8 is observed and the yield of the protein increases ~2 times.

Heme is important for the pri-miRNA processing activity of DGCR8

Like several other heme-binding proteins that are also discovered serendipitously, we used biochemical assays to characterize DGCR8, focusing on two major questions: how does heme affect its pri-miRNA processing activity and how does it interact with heme. The activity of recombinant DGCR8 proteins were tested using reconstituted pri-miRNA processing assays. In these assays, fragments of pri-miRNAs are labeled with ^{32}P and are incubated with purified recombinant Drosha and DGCR8. The reactions are analyzed using denaturing polyacrylamide gel electrophoresis and autoradiography. A pri-miRNA is typically cleaved into three products, with the one in the middle being the pre-miRNA (Fig. 1). We found that the heme-bound DGCR8 is more active than the heme-free form in pri-miRNA processing *in vitro* (25). This is true for both pri-miRNAs we tested. Thus, the heme-bound NC1 is an active form of DGCR8 and is unlikely an artifact of overexpression of a human protein in *E. coli*. Furthermore, DGCR8 binds heme using an independently folded heme-binding domain (HBD) located in the central region of its sequence (28). The presence and functional importance of heme suggest a mechanism for regulation of DGCR8.

Redox regulation of pri-miRNA processing activity of heme-bound DGCR8 by glutathione

A class of heme-binding proteins serves as sensors for diatomic gas molecules, such as nitric oxide (NO) and carbon monoxide (CO), or for redox potential (29). We set out to identify the function of the heme-DGCR8 interaction using the reconstituted pri-miRNA processing assays. We found that the heme-bound DGCR8 protein is activated by reduced glutathione and is inhibited by oxidized glutathione (Senturia *et al.* submitted). Glutathione is a natural tripeptide (γ-glutamyl-cysteinyl-glycine) involved in oxidation-reduction reactions and signaling (30). Further characterization of this regulation revealed the following features. (1) The effect of glutathione on pri-miRNA processing is achieved specifically through DGCR8 because the reduced and oxidized glutathione can be removed through buffer exchange after pre-incubation with the heme-bound NC1 and DGCR8 remains activated or inhibited, respectively. (2) The bidirectional regulation of DGCR8 by glutathione is reversible. (3) The association of DGCR8 with heme is required for the sensitivity to glutathione: two heme-free DGCR8 proteins failed to respond to glutathione. The first (called NC9) does not contain the HBD (Fig. 2) (25) and the second contains a mutation (P351A) in the HBD. (4) Although a disulfide bond between the HBD and the dsRBD1 can be detected, disruption of this interaction through a cysteine-to-serine mutation does not affect the sensitivity to glutathione. This result, along with other evidence, indicates that the effect of glutathione on DGCR8 is unlikely mediated by a simple mechanism depending on a disulfide. Furthermore, our biochemical assays have largely ruled out the possibility of DGCR8 acting as a sensor for diatomic gas molecules, such as nitric oxide (NO) and carbon monoxide (CO). Thus, this study suggests that a primary function of the heme–DGCR8 interaction is for sensing glutathione-mediated redox potential in cells, which could fluctuate during development in oxidative stress or hypoxia conditions. We are in the process of testing whether miRNA maturation is affected by redox potential in cultured mammalian cells.

A novel use of WW motif for dimerization and for association with heme

The DGCR8 protein represents a novel family of heme-binding proteins. The only recognizable motif in its heme-binding domain is a WW motif located in its N-terminal region (Fig. 2). The WW motifs typically contain ~30 amino acids with two characteristic tryptophan residues and are widely distributed in protein families (31). Many WW-motif-containing proteins function in signaling pathways. Most WW motifs that have been characterized interact with proline-containing sequences.

An important feature of the DGCR8–heme interaction is that the dimeric form of DGCR8 binds heme, whereas monomers do not. In an attempt to isolate mutations that abolish heme binding in DGCR8, we identified a conserved dimerization domain in DGCR8. It is embedded in the N-terminal region of the HBD, including the WW motif and its C-terminal neighboring region, as well as some residues on its N-terminal side (Fig. 2). The crystal structure of the dimerization domain was determined at 1.7 Å resolution (Fig. 3). The structure reveals that the WW motif of DGCR8 forms a 3-stranded β-sheet, which is superimposible to other known WW domain structures. More importantly, each WW motif makes extensive contacts with the C-terminal neighboring region of the partner subunit in a dimer. This exchange of interacting partners between subunits in an oligomeric protein is called three-dimensional domain swapping (32). The surface that a classic WW domain uses to interact with proline-containing ligands is adopted by DGCR8 as a part of the dimerization interface. Thus, it is unlikely that the WW motif in DGCR8 binds proline-rich ligands.

Although the structure of the DGCR8 dimerization domain does not contain heme, it does contain two residues that we have already identified through mutagenesis studies to be required for association with heme. Cys352 is completely conserved in the DGCR8 family. Mutation of this residue to either alanine, serine or histidine completely abolishes heme-binding to NC1 (25). Because only the heme cofactors with thiolate ligands have been shown to have Soret peaks at 450-nm wavelength in their

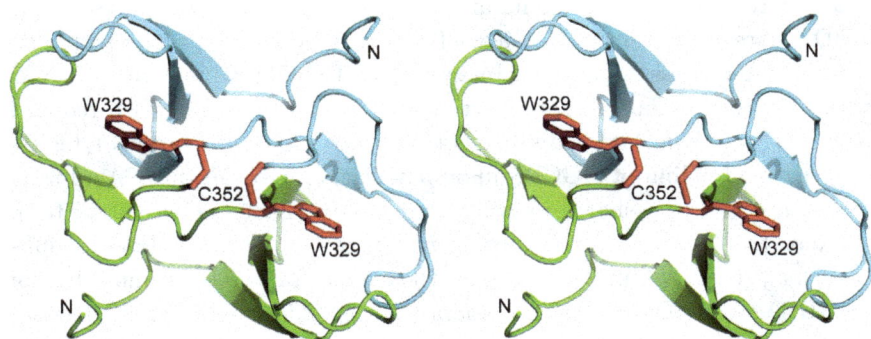

Fig. 3. The dimerization domain of DGCR8 contributes a heme-binding surface. In the stereo diagram, the two subunits of DGCR8 are drawn in green and cyan and the residues required for association with heme in red.

UV-Vis absorption spectra, such as those in the cytochrome P450 proteins, it is likely that the Cys352 side chain serves as the coaxial ligand of the heme bound to DGCR8. The second residue essential for heme binding is Trp329, which is the second tryptophan in the WW motif (28). Interestingly, Cys352 and Trp329 contributed by both DGCR8 subunits cluster on an elongated surface of the dimerization domain (Fig. 3). Our structure and mutagenesis data suggest that the dimerization domain directly contributes a surface for binding heme and explains why only dimeric DGCR8 proteins bind heme. Thus, we conclude that DGCR8 makes a novel use of a WW motif as a structural platform for dimerization and for association with heme.

The heme-binding-deficient DGCR8 mutants, including both C352A and W329A, have reduced pri-miRNA processing activity *in vitro*, consistent with a regulatory function of the HBD and heme in pri-miRNA processing (25, 28).

The C-terminal region of the HBD is an autoinhibitory domain

Using an anti-DGCR8 antibody, we found that DGCR8 undergoes two proteolytic cleavages in HeLa cells (Gong *et al.* submitted). The C-terminal cleavage fragments of DGCR8, called DGCR8^{C1} and DGCR8^{C2}, can be observed on immunoblots. The two cleavage sites (CS1 and CS2) have been mapped to the immediate N-terminal neighboring region of the HBD and in the middle of the HBD, respectively. We further identified caspases as mediating DGCR8 cleavage at CS2. The cleavage of recombinant DGCR8 results in loss of heme and dissociation of the two halves of the HBD. More importantly, cleavage of DGCR8 by caspases inactivates its pri-miRNA processing activity. Interestingly, DGCR8^{C2} binds pri-miRNAs with affinities similar to that of active DGCR8 proteins, but with reduced cooperativity. The latter finding corroborates the importance of highly cooperative binding of DGCR8 upon association with pri-miRNAs during processing we have shown in a separate study (Faller *et al.* submitted). Because the C-terminal neighboring region of the DGCR8 HBD is sufficient for pri-miRNA processing, our study identifies the C-terminal half of the HBD as an autoinhibition domain (Fig. 4). The molecular mechanism revealed here provides an interesting example of how proteases inactivate heme-binding proteins through disassembly of their heme-binding domains.

Our characterization of DGCR8 revealed a domain structure that appears to be responsible for regulating its pri-miRNA processing activity

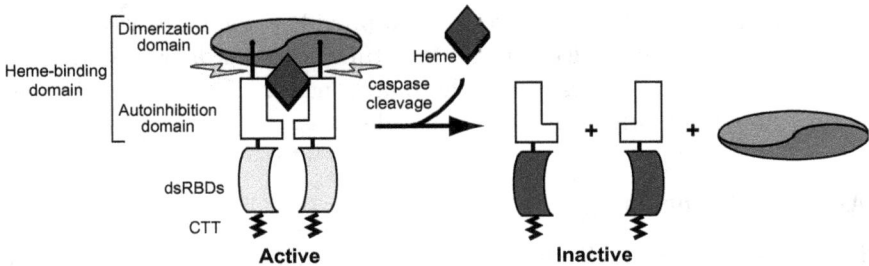

Fig. 4.   Caspase cleavage of DGCR8 results in loss of heme and inactivation of its pri-miRNA processing activity.

(Fig. 2). The double-stranded RNA-binding domains (dsRBDs) and C-terminal tail (CTT) in the C-terminal region of DGCR8 are sufficient for pri-miRNA binding and processing, and thus are the effector domains. The N-terminal region of DGCR8 (residues 1–275) is required for its nuclear localization (27). The central region is the HBD, which contains the dimerization and autoinhibition domains separated by the central loop. The autoinhibition domain directly interacts with the dsRBDs and can inhibit the pri-miRNA binding and processing activity. Thus, the dimerization domain is effectively responsible for activation of the pri-miRNA processing activity of DGCR8. It likely does so through a direct interaction with the autoinhibition domain and the heme cofactor.

Future directions

Our series of biochemical and structural studies demonstrate intrinsic, conserved properties of the DGCR8 protein: association with heme, regulation by glutathione in a heme-dependent manner, and the modular architecture of its heme-binding domain. We and our collaborators are working on answering the following questions: Does DGCR8 bind heme in human or other eukaryotic cells? How do oxidative stress and other conditions that perturb the balance between reduced and oxidized glutathione affect miRNA biogenesis in cells and animals? From the perspective of fundamental protein chemistry, how does DGCR8 bind heme with its unique characteristics, such as the 450-nm absorption peak in the absence of the obvious source of carbon monoxide? How do reduced and oxidized glutathione interact with the heme-bound DGCR8 and regulate its activity?

Through these investigations, we hope to provide a comprehensive understanding, at cellular, molecular and structural levels, of how miRNA processing is regulated. This knowledge may suggest ways to explain and correct abnormal miRNA maturation in diseases.

Acknowledgements

I would like to thank Rachel Senturia for comments on the manuscript and for assistance with the figure preparation, and other members of my laboratory for discussions.

References

1. Bushati N, Cohen SM. 2007. microRNA functions. *Annu Rev Cell Dev Biol* 23: 175–205.
2. Bartel DP. 2009. MicroRNAs: Target recognition and regulatory functions. *Cell* 136: 215–233.
3. Filipowicz W, Bhattacharyya SN, Sonenberg N. 2008. Mechanisms of post-transcriptional regulation by microRNAs: Are the answers in sight? *Nat Rev Genet* 9: 102–114.
4. Hou W, Tian Q, Zheng J, Bonkovsky HL. 2010. MicroRNA-196 represses Bach1 protein and hepatitis C virus gene expression in human hepatoma cells expressing hepatitis C viral proteins. *Hepatology* 51: 1494–1504.
5. Igarashi K, Sun J. 2006. The heme-Bach1 pathway in the regulation of oxidative stress response and erythroid differentiation. *Antioxid Redox Signal* 8: 107–118.
6. Kim VN, Han J, Siomi MC. 2009. Biogenesis of small RNAs in animals. *Nat Rev Mol Cell Biol* 10: 126–139.
7. Faller M, Guo F. 2008. MicroRNA biogenesis: There's more than one way to skin a cat. *Biochim Biophys Acta* 1779: 663–667.
8. Lee Y, Ahn C, Han J, Choi H, Kim J, Yim J, Lee J, Provost P, Radmark O, Kim S, Kim VN. 2003. The nuclear RNase III Drosha initiates microRNA processing. *Nature* 425: 415–419.
9. Denli AM, Tops BB, Plasterk RH, Ketting RF, Hannon GJ. 2004. Processing of primary microRNAs by the Microprocessor complex. *Nature* 432: 231–235.
10. Gregory RI, Yan KP, Amuthan G, Chendrimada T, Doratotaj B, Cooch N, Shiekhattar R. 2004. The Microprocessor complex mediates the genesis of microRNAs. *Nature* 432: 235–240.

11. Han J, Lee Y, Yeom KH, Kim YK, Jin H, Kim VN. 2004. The Drosha-DGCR8 complex in primary microRNA processing. *Genes Dev* 18: 3016–3027.

12. Landthaler M, Yalcin A, Tuschl T. 2004. The human DiGeorge syndrome critical region gene 8 and its D. melanogaster homolog are required for miRNA biogenesis. *Curr Biol* 14: 2162–2167.

13. Hammond SM, Boettcher S, Caudy AA, Kobayashi R, Hannon GJ. 2001. Argonaute2, a link between genetic and biochemical analyses of RNAi. *Science* 293: 1146–1150.

14. Mourelatos Z, Dostie J, Paushkin S, Sharma A, Charroux B, Abel L, Rappsilber J, Mann M, Dreyfuss G. 2002. miRNPs: A novel class of ribonucleoproteins containing numerous microRNAs. *Genes Dev* 16: 720–728.

15. Croce CM. 2009. Causes and consequences of microRNA dysregulation in cancer. *Nat Rev Genet* 10: 704–714.

16. Lu J, Getz G, Miska EA, Alvarez-Saavedra E, Lamb J, Peck D, Sweet-Cordero A, Ebert BL, Mak RH, Ferrando AA, Downing JR, Jacks T, Horvitz HR, Golub TR. 2005. MicroRNA expression profiles classify human cancers. *Nature* 435: 834–838.

17. Thomson JM, Newman M, Parker JS, Morin-Kensicki EM, Wright T, Hammond SM. 2006. Extensive post-transcriptional regulation of microRNAs and its implications for cancer. *Genes Dev* 20: 2202–2207.

18. Kumar MS, Lu J, Mercer KL, Golub TR, Jacks T. 2007. Impaired microRNA processing enhances cellular transformation and tumorigenesis. *Nat Genet* 39: 673–677.

19. Wang Y, Medvid R, Melton C, Jaenisch R, Blelloch R. 2007. DGCR8 is essential for microRNA biogenesis and silencing of embryonic stem cell self-renewal. *Nat Genet* 39: 380–385.

20. Yi R, Pasolli HA, Landthaler M, Hafner M, Ojo T, Sheridan R, Sander C, O'Carroll D, Stoffel M, Tuschl T, Fuchs E. 2009. DGCR8-dependent microRNA biogenesis is essential for skin development. *Proc Natl Acad Sci U S A* 106: 498–502.

21. Shiohama A, Sasaki T, Noda S, Minoshima S, Shimizu N. 2003. Molecular cloning and expression analysis of a novel gene DGCR8 located in the DiGeorge syndrome chromosomal region. *Biochem Biophys Res Commun* 304: 184–190.

22. Han J, Pedersen JS, Kwon SC, Belair CD, Kim YK, Yeom KH, Yang WY, Haussler D, Blelloch R, Kim VN. 2009. Posttranscriptional crossregulation between Drosha and DGCR8. *Cell* 136: 75–84.

23. Triboulet R, Chang HM, Lapierre RJ, Gregory RI. 2009. Post-transcriptional control of DGCR8 expression by the Microprocessor. *RNA* 15: 1005–1011.

24. Han J, Lee Y, Yeom KH, Nam JW, Heo I, Rhee JK, Sohn SY, Cho Y, Zhang BT, Kim VN. 2006. Molecular basis for the recognition of primary microRNAs by the Drosha-DGCR8 complex. *Cell* 125: 887–901.

25. Faller M, Matsunaga M, Yin S, Loo JA, Guo F. 2007. Heme is involved in microRNA processing. *Nat Struct Mol Biol* 14: 23–29.

26. Sohn SY, Bae WJ, Kim JJ, Yeom KH, Kim VN, Cho Y. 2007. Crystal structure of human DGCR8 core. *Nat Struct Mol Biol* 14: 847–853.

27. Yeom KH, Lee Y, Han J, Suh MR, Kim VN. 2006. Characterization of DGCR8/Pasha, the essential cofactor for Drosha in primary miRNA processing. *Nucleic Acids Res* 34: 4622–4629.

28. Senturia R, Faller M, Yin S, Loo JA, Cascio D, Sawaya MR, Hwang D, Clubb RT, Guo F. 2010. Structure of the dimerization domain of DiGeorge Critical Region 8. *Protein Sci* 19: 1354–1365.

29. Gilles-Gonzalez MA, Gonzalez G. 2005. Heme-based sensors: Defining characteristics, recent developments, and regulatory hypotheses. *J Inorg Biochem* 99: 1–22.

30. Biswas SK, Rahman I. 2009. Environmental toxicity, redox signaling and lung inflammation: The role of glutathione. *Mol Aspects Med* 30: 60–76.

31. Ilsley JL, Sudol M, Winder SJ. 2002. The WW domain: Linking cell signalling to the membrane cytoskeleton. *Cell Signal* 14: 183–189.

32. Liu Y, Eisenberg D. 2002. 3D domain swapping: As domains continue to swap. *Protein Sci* 11: 1285–1299.

Note added in proof: The submitted manuscript by Faller *et al.* has been published in *RNA* 16: 1570–1583 in 2010.

Chapter 7

THE VAST POTENTIAL OF HEME IN REGULATING BIOLOGICAL PROCESSES

A Global Perspective

Li Zhang, Aaron Arvey, Donovan Pham Huynh and Christina Leslie

Overview

Clinical and experimental data have clearly shown that heme is crucial for proper cell functioning and for the survival and development of humans. In previous chapters, we have described the diseases caused by heme deficiency, and have provided examples of how heme can control gene transcription, protein synthesis, and microRNA biogenesis. Here, we further examine the potentially broad role of heme in diverse molecular and cellular processes underlying proper human functioning and disease pathogenesis.

Molecular and cellular events underlie proper human functioning

Life is a fascinating and miraculous form of existence in the universe. The many capabilities afforded by living organisms, particularly humans, are still impossible to match by engineering, and will remain so for a long time. How can humans act in such rapid, coordinated, and sophisticated manners, for example, in the fight-or-flight response when facing danger or in various sporting activities? Still, when under attack due to internal problems or external infections or injuries, humans can fail rapidly and completely. How are such dramatic responses and changes brought about?

Fortunately, the advances in biological and biomedical sciences have allowed us to glean the seemingly invisible events that underlie the behaviors and responses of living organisms and humans. It turns out that for every movement or response by an organism or human, there are biophysical and biochemical events preceding it. Humans are able to respond to danger because of the actions of the molecules and cells in various parts of our body, starting from the eyes to the nerve cells and muscle cells. We may become ill because the molecules and cells in our body do not act properly.

Heme controls the growth, differentiation, and death of diverse cells

Remarkably, heme is a molecule that can act throughout the human body to control and regulate diverse cells and processes (1–3). Heme appears to play important regulatory roles in virtually all human or mammalian cells, which several laboratories have examined to date (4–8). Heme has been shown to be critical for mammalian cell growth. For example, lack of heme synthesis causes the human epithelial cervix carcinoma HeLa cells to stop growth and undergo programmed cell death or apoptosis (7, 8). Figure 1 shows that HeLa cells treated with succinyl acetone, a potent inhibitor of heme synthesis, undergo DNA fragmentation, and programmed cell death (apoptosis), as detected by the TUNNEL assay. Likewise, neuronal cells undergo cell death under heme-deficient conditions (9) (see Fig. 2). Heme also promotes the differentiation of mammalian erythroid, adipose, and neuronal cells (2, 4, 10). Such effects of heme are mediated by the actions of various regulatory proteins. Some of these proteins include those described in detail in Chapters 3–5, for example, Bach1 and HRI. They bind to heme directly and respond to changes of heme concentration.

Furthermore, heme can affect the activities of many key proteins controlling cell function, growth, and death, even when heme may not bind to them directly (1, 3, 11–15). For example, work in my lab showed that heme deficiency increases the protein levels of the tumor suppressor gene product p53 and CDK inhibitor p21, and decreases the protein levels of Cdk4, Cdc2, and cyclin D2 (7). In addition, we showed that heme

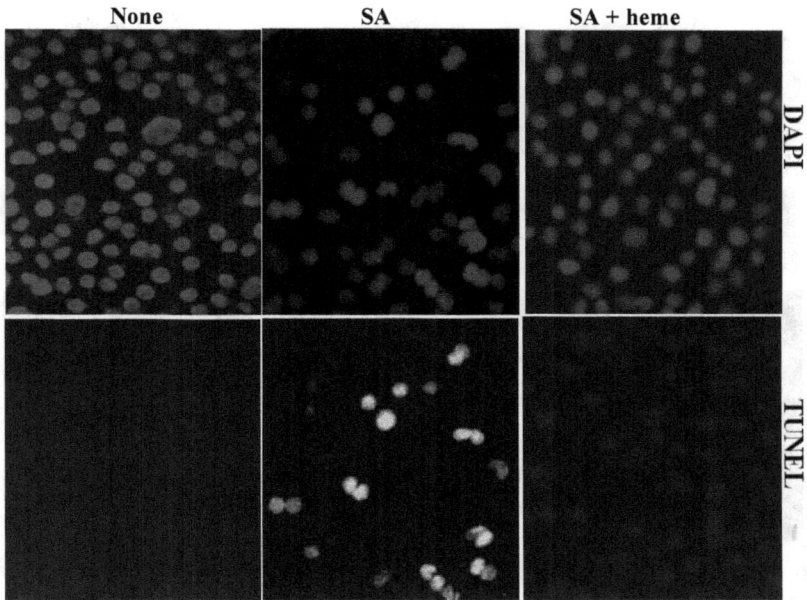

Fig. 1. Inhibition of heme synthesis in HeLa cells causes the cells to undergo programmed cell death. Shown here are the untreated HeLa cells (None), cells treated with succinyl acetone, a potent inhibitor of heme synthesis (SA), and cells treated with both succinyl acetone and heme (SA + heme). Heme addition reverses the effect of succinyl acetone on cell death. Upper panel: DAPI images showing the nuclei of HeLa cells. Lower panel: Fluorescent TUNEL (Terminal deoxynucleotidyl transferase dUTP nick end labeling) images showing cells undergoing program cell death. The TUNEL assay detects DNA fragmentation in apoptotic cells. Cells treated with succinyl acetone are apoptotic.

deficiency diminishes the activation/phosphorylation of Raf, MEK1/2, and ERK1/2 components of the mitogen-activated protein (MAP) kinase signaling pathway, in both the human epithelial cervix carcinoma HeLa cells and in the nerve growth factor-induced rat pheochromocytoma PC12 cells (6, 7). Heme deficiency also directly inhibits the function of the glutamate NMDA receptor and causes neurite damage in primary mouse cortical neurons (16). These effects of heme deficiency on neuronal cells provide a sound explanation for the neurological disturbances associated with acute porphyrias, described in Chapter 2.

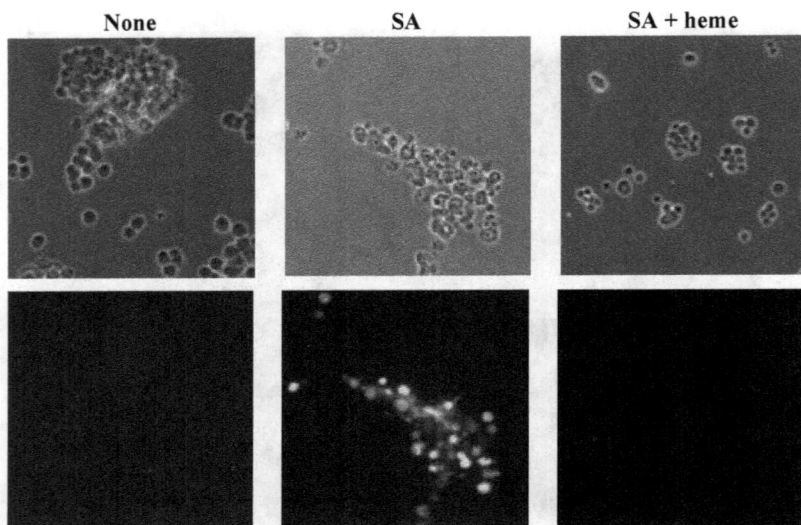

Fig. 2. Inhibition of heme synthesis in PC12 cells causes the cells to undergo programmed cell death. PC12 cells induced with nerve growth factor treated with no reagent (None), succinyl acetone (SA), and succinyl acetone and heme (SA + heme) were collected and subjected to TUNEL assays. Upper panel: The bright field images of cells used for TUNEL assays. Lower panel: Fluorescent TUNEL images showing cells undergoing program cell death. Cells treated with succinyl acetone are apoptotic.

Heme has the potential to modulate the activities of many regulatory proteins that are crucial for proper human functioning

Importantly, besides those regulatory proteins described in previous chapters, emerging evidence suggests that heme has the potential to control the function of many other proteins (17). Our recent computational study shows that 5322 human proteins (out of a total of 20 495) contain one or more heme-binding CP motifs or heme-responsive or regulatory motifs (HRMs), as described in previous chapters (17). Most of the known heme-regulated proteins that bind to heme and respond to changes of heme concentration contain one or more CP motifs (3). In addition, mutations in the CP motifs of Hap1, Bach1, and IRR abolish heme regulation of their activity (18–24). Therefore, the existence of CP motifs in proteins is a good indicator for the potential of proteins to bind heme reversibly and to be regulated by heme concentration.

The proteins containing CP motifs fall into highly statistically significant functional categories. Table 1 lists some of these functional categories. For example, 92 out of 158 protein tyrosine kinases and 31 out of 49 proteins involved in small GTPase-mediated signal transduction

Table 1. The numbers of proteins with CP motifs in selected functional categories.

Protein functional category	The number of proteins with CP[a]	Total number of human proteins[b]	P-value
Protein tyrosine kinase activity	92	158	0
Small GTPase mediated signal transduction	31	49	4.32E-08
Chromatin modification	75	146	1.06E-11
Nucleic acid binding	502	1516	3.00E-10
Transcription activator activity	58	151	0.00051167
Transcription coactivator activity	61	164	0.00097419
Steroid hormone receptor activity	26	49	4.75E-05
Ephrin receptor activity	14	14	6.31E-09
Ephrin receptor binding	6	6	0.00030595
Brain development	32	74	0.00092149
Metabotropic glutamate receptor activity	17	23	2.05E-06
Gamma-aminobutyric acid signaling	11	15	0.00016629
Voltage-gated ion channel activity	81	151	0
Proteinaceous extracellular matrix	135	277	1.62E-10
Ubiquitin-protein ligase activity	57	127	2.99E-06
Metal ion binding	1099	3060	0
Zinc ion binding	922	2439	1.35E-10
Calcium ion binding	324	878	5.28E-10
Iron-sulfur cluster binding	24	46	0.00013089

[a] The numbers of human proteins containing one or more potential heme-binding CP motifs in each functional category; [b] The total numbers of human proteins in each functional category.

contain CP motifs. These tyrosine kinases and small GTPases are involved in most, if not all, cellular responses to extracellular stimuli, and are crucial for the control of cell growth, differentiation, and death (25–29). Their dysfunction is associated with numerous diseases including many cancers. Likewise, many regulators of gene expression, including proteins involved in chromatin modification, transcription activators and coactivators, and steroid hormone receptors, contain CP motifs (Table 1) (30–32).

Intriguingly, these proteins include many involved in human development and brain development. Particularly, all ephrin receptors and receptor-binding proteins, which are involved in development and are associated with many diseases including cancers, contain CP motifs (Table 1) (33–36). Although the mere presence of CP motifs does not necessarily suggest functional importance or heme regulation, this high frequency of CP motif occurrences in these proteins is of great statistical significance. Many neuronal proteins, such as those involved in the metabotropic glutamate and gamma-aminobutyric acid signaling pathway (37–40), contain CP motifs (Table 1). This agrees very well with recent experimental studies showing that heme is important for the proper functioning of NMDA receptor and large conductance Ca^{2+}- and voltage-activated Slo1 K^+ (41, 42).

To gain further insights into the significance of these human proteins containing CP motifs, we used a computational algorithm called Pathway Studio (Ariadne) to find the connections between these proteins. Strikingly, we found that 253 of the proteins containing CP motifs are connected to the epidermal growth factor receptor (EGFR) (Fig. 3). In other words, these 253 proteins modulate or are modulated by the EGF receptor; they work with the receptor in controlling many processes in cells. Defects in the EGF receptor function are associated with many cancers, including non-small-cell lung cancer, glioblastoma, breast cancer, and colorectal cancer (43–49). Likewise, 117 proteins that are in the ERBB2 regulatory network contain one or more CP motifs (Fig. 4). ERBB2 is the v-erb-b2 erythroblastic leukemia viral oncogene homolog 2, a homolog of the EGF receptor (50–52). Defects in ERBB2 function are also associated with many cancers, such as breast cancer (53–55). Notably, 109 proteins in the insulin-like growth factor 1 receptor (IGF1R) regulatory network contain one or more CP motifs (Fig. 5) (56, 57). Defects in IGF1R may result in intrauterine and post-natal growth retardation, type 2 diabetes, and cancer (58–64). Recent

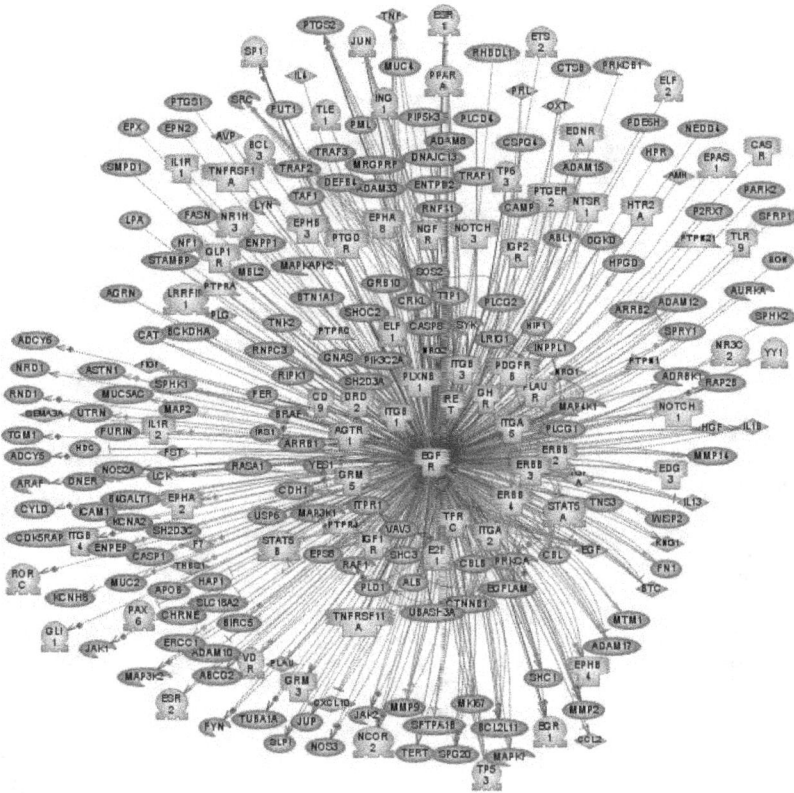

Fig. 3. The epidermal growth factor receptor (EGFR) regulatory network formed by proteins containing one or more CP motifs.

experimental evidence in Dr. Zhang's laboratory suggests that heme may indeed control the activity of some of these proteins containing one or more CP motifs.

Experimental evidence shows that heme can indeed bind to and regulate the activity of protein tyrosine kinases

Protein tyrosine kinases play crucial roles in signal transduction in virtually all mammalian cells (25, 65–67). They are enzymes that phosphorylate other proteins and enzymes, in response to a signal or stimulus. By phosphorylating other proteins and enzymes, they modulate the activity

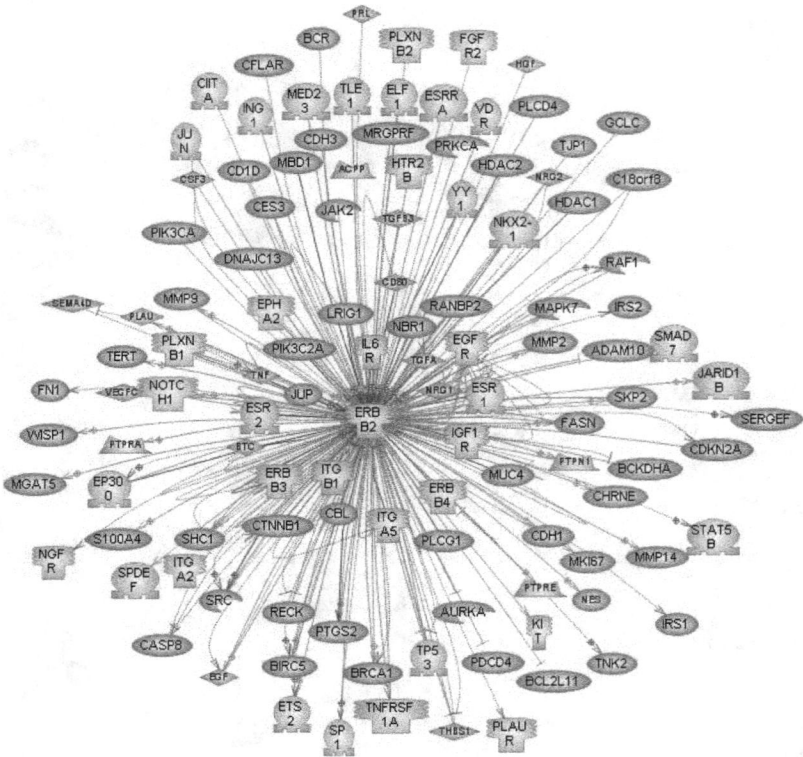

Fig. 4. The ERBB2 regulatory network formed by proteins containing one or more CP motifs.

of diverse cellular signaling and regulatory proteins. They coordinate cellular responses to a wide array of extracellular stimuli, and regulate cell growth, survival, differentiation, migration, and metabolism (68–71) (see Figs. 6 and 7 for examples). Deregulated expression or activity of tyrosine kinases can promote serious diseases, particularly cancers (72–81). As such, understanding the molecular mechanisms by which the activity of tyrosine kinases is controlled is a key subject in biological and biomedical research.

Interestingly, our recent computational analysis found that a very high fraction of protein tyrosine kinases contain one or more CP motifs (see Table 1). Particularly, 23 non-membrane spanning or non-receptor tyrosine kinases out of 39 known tyrosine kinases (p value: 1.33E-05;

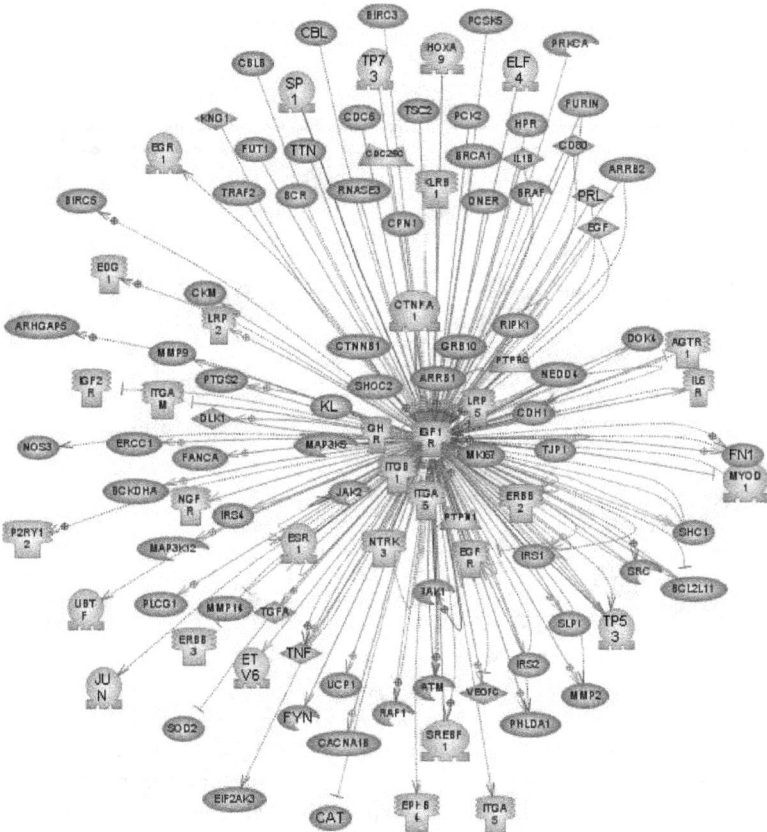

Fig. 5. The insulin-like growth factor 1 receptor (IGF1R) regulatory network formed by proteins containing one or more CP motifs.

p < 0.000527 after correction for multiple tests) contain at least one CP motif (17). We were able to closely examine the effect of heme on two important tyrosine kinases, Jak2 and Src (Figs. 8 and 9), *in vivo* and *in vitro*. *In vivo*, we found that heme can modulate the phosphorylation state of Jak2 at Tyr1007/1008 and of Src at Tyr530 (Figs. 8 and 9). The phosphorylation states of Jak2 and Src at these residues are crucial for determining the activities of Jak2 and Src in diverse cells. *In vitro*, we found that heme can bind to the purified, full-length Src protein and the Jak2 fragment containing residues 808–1132. In addition, heme binding alters

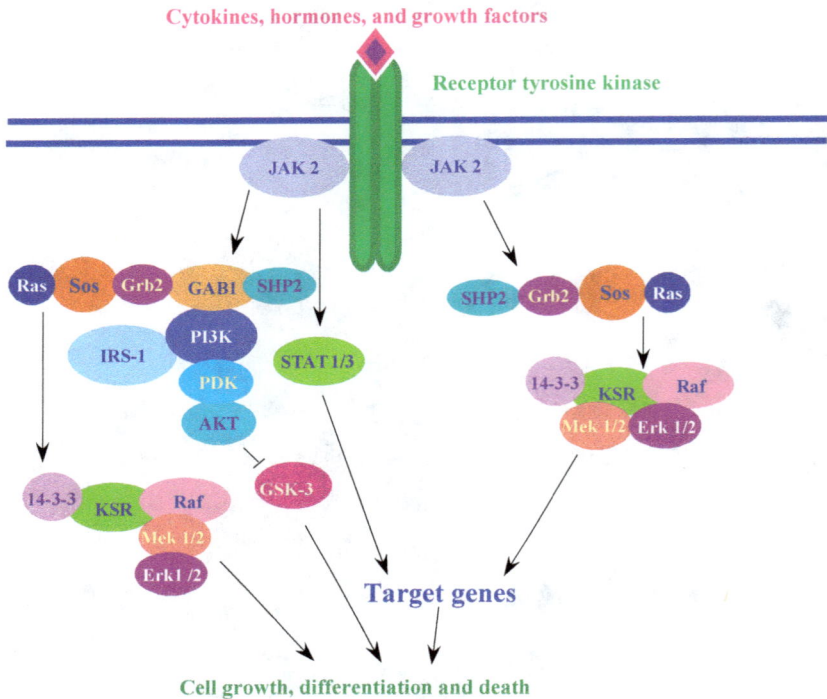

Fig. 6. Examples of signaling pathways involving Jak2. Shown here are several Jak2-mediated signaling cascades that enable cells to respond extracellular signals or stimuli and initiate changes in gene expression, cell growth, differentiation and death. Abbreviations: RTK: receptor tyrosine kinase; Jak2: janus kinase 2; SHP2 : tyrosine phosphatase non-receptor type 11; Grb2: growth factor receptor-bound protein 2; Sos: son of sevenless homolog 2; Ras: rat sarcoma; Mek: MAP kinase; Erk: extracellular signal-regulated kinases; KSR: kinase suppressor of Ras; GSK-3β: glycogen synthase kinase-3 beta: STAT: signal transducer and activator of transcription; GAB1: GRB2-associated binding protein 1; PI3K: phosphatidylinositol-4,5-bisphosphate 3 kinase; PDK: pyruvate dehydrogenase kinase isozyme; IRS1: insulin receptor substrate 1.

the sensitivity of Jak2 and Src to proteases, suggesting that heme binds to these proteins and alters their conformation.

The regulation of Jak2 and Src by heme may have profound implications in understanding numerous biological processes and in the management and treatment of numerous diseases. Jak2 is one of the key protein

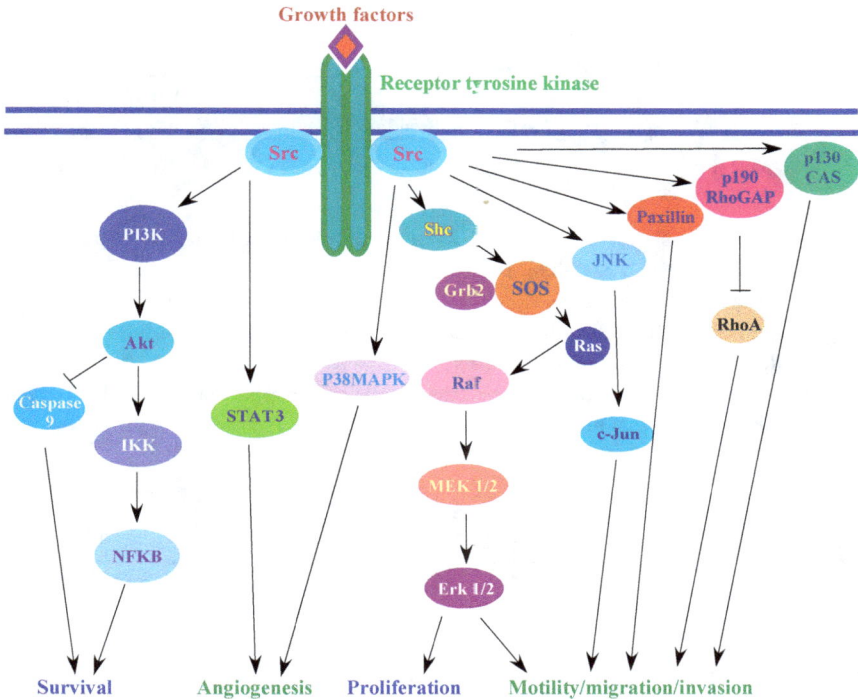

Fig. 7. Examples of signaling pathways involving Src. Shown here are several Src-mediated signaling cascades that enable cells to respond growth factors and promote changes in angiogenesis, cell proliferation, survival, motility, migration, and invasion. Abbreviations are as in Fig. 6.

tyrosine kinases that mediate cellular signaling by a variety of cytokines, hormones and growth factors (72, 76, 80, 82–95). Jak2 is essential for erythropoietin receptor (EpoR) signal transduction (96–98). Jak2 knockout mice exhibit embryonic lethality, and no erythropoiesis occurs in these mice (96, 97). Mutations in Jak2 have been found to be associated with an array of hematopoietic diseases, including myelofibrosis, myeloproliferative disorders, and polycythemia vera (99–109). Particularly, the V617F Jak2 mutation is present in 90%–95% of patients with polycythemia vera, 50%–70% of patients with essential thrombocythemia, and 40%–50% of patients with primary myelofibrosis (110). The role of heme in stimulating the phosphorylation of Jak2 and its activity is in perfect agreement with the role of Jak2 in erythropoiesis, because heme is arguably the

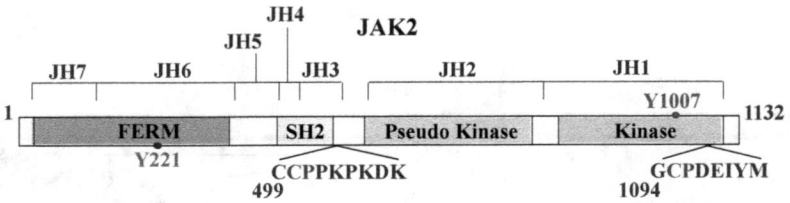

Fig. 8. Domain structure of Jak2. Jak2 contains the FERM (four-point-one, ezrin, radixin, moesin), SH2-like, pseudokinase and kinase domains. An alternative nomenclature for the putative domains is as a series of the Janus homology (JH) domains. The Tyr1007 phosphorylation site and the CP motifs are indicated. The FERM domain mediates binding to cytokine receptors. Both the FERM and the pseudokinase domains regulate catalytic activity. Phosphorylation of Tyr1007 is critical for Jak2 activation.

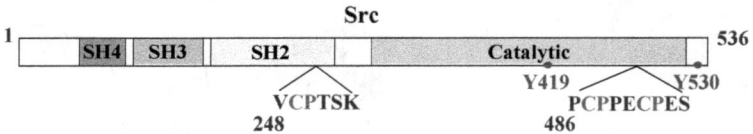

Fig. 9. Domain structure of Src. The domains shown are the Src homology (SH) domains, the catalytic domain, the unique SH4 domain, and SH2 and SH3 domains. Phosphorylation of the Tyr530 residue inhibits Src activity. Phosphorylation of the Tyr419 residue activates Src activity. The CP motifs are indicated.

most crucial molecule in erythropoiesis. As mentioned previously, 80% of the heme in the human body is devoted to the generation of red blood cells (111).

Src is the first discovered oncogene, and it is one of the best studied. Src is responsible for signal transduction in controlling many cellular processes, including cell growth, differentiation, adhesion, invasion and metastasis (73–75, 112–124). Increased Src activity frequently occurs in many types of human cancer. It is a key molecule in promoting tumor progression. It has therefore emerged as a potential therapeutic target for treating many types of human cancer, including breast cancer, prostate cancer, lung cancer and brain cancer (73, 112, 115–117, 121, 124). It is worth noting that heme can promote the phosphorylation of the Tyr530 residue (17), which plays a critical role in regulating the kinase activity of Src (see Fig. 9). This raises the possibility that heme and analogs have the potential to modulate the progression of an array of human cancers.

These examples shed light on the potentially broad role of heme in proper cell functioning and in the development and health of humans. If heme can control such important proteins crucial for many physiological and disease processes, then the roles of heme may be exploited in treating various diseases such as cancers. As more is known about how heme controls the activities of proteins such as tyrosine kinases, heme analogs may be exploited as drugs for many diseases. Indeed, in Chapter 9, Dr. Gao and colleagues explain how heme and heme precursors can be used to treat cancers. However, this application is not based on the regulatory role of heme. The potentially broad regulatory role of heme can be further exploited in medicine.

References

1. Hou S, Reynolds MF, Horrigan FT, Heinemann SH, Hoshi T. 2006. Reversible binding of heme to proteins in cellular signal transduction. *Acc Chem Res* 39: 918–924.

2. Padmanaban G, Venkateswar V, Rangarajan PN. 1989. Haem as a multifunctional regulator. *Trends Biochem Sci* 14: 492–496.

3. Mense SM, Zhang L. 2006. Heme: A versatile signaling molecule controlling the activities of diverse regulators ranging from transcription factors to MAP kinases. *Cell Res* 16: 681–692.

4. Rutherford TR, Clegg JB, Weatherall DJ. 1979. K562 human leukaemic cells synthesise embryonic haemoglobin in response to haemin. *Nature* 280: 164–165.

5. Zhu Y, Hon T, Zhang L. 1999. Heme initiates changes in the expression of a wide array of genes during the early erythroid differentiation stage. *Biochem Biophys Res Commun* 258: 87–93.

6. Zhu Y, Hon T, Ye W, Zhang L. 2002. Heme deficiency interferes with the Ras-mitogen-activated protein kinase signaling pathway and expression of a subset of neuronal genes. *Cell Growth Differ* 13: 431–439.

7. Ye W, Zhang L. 2004. Heme controls the expression of cell cycle regulators and cell growth in HeLa cells. *Biochem Biophys Res Commun* 315: 546–554.

8. Ye W, Zhang L. 2004. Heme deficiency causes apoptosis but does not increase ROS generation in HeLa cells. *Biochem Biophys Res Commun* 319: 1065–1071.

9. Sengupta A, Hon T, Zhang L. 2005. Heme deficiency suppresses the expression of key neuronal genes and causes neuronal cell death. *Brain Res Mol Brain Res* 137: 23–30.

10. Ishii DN, Maniatis GM. 1978. Haemin promotes rapid neurite outgrowth in cultured mouse neuroblastoma cells. *Nature* 274: 372–374.

11. Tsiftsoglou AS, Tsamadou AI, Papadopoulou LC. 2006. Heme as key regulator of major mammalian cellular functions: Molecular, cellular, and pharmacological aspects. *Pharmacol Ther* 111: 327–345.

12. Weeks CL, Singh S, Madzelan P, Banerjee R, Spiro TG. 2009. Heme regulation of human cystathionine beta-synthase activity: Insights from fluorescence and Raman spectroscopy. *J Am Chem Soc* 131: 12809–12816.

13. Yi L, Jenkins PM, Leichert LI, Jakob U, Martens JR, Ragsdale SW. 2009. Heme regulatory motifs in heme oxygenase-2 form a thiol/disulfide redox switch that responds to the cellular redox state. *J Biol Chem* 284: 20556–20561.

14. Faller M, Matsunaga M, Yin S, Loo JA, Guo F. 2007. Heme is involved in microRNA processing. *Nat Struct Mol Biol* 14: 23–29.

15. Wang H, Zhou Q, Kesinger JW, Norris C, Valdez C. 2007. Heme regulates exocrine peptidase precursor genes in zebrafish. *Exp Biol Med (Maywood)* 232: 1170–1180.

16. Chernova T, Steinert JR, Guerin CJ, Nicotera P, Forsythe ID, Smith AG. 2007. Neurite degeneration induced by heme deficiency mediated via inhibition of NMDA receptor-dependent extracellular signal-regulated kinase 1/2 activation. *J Neurosci* 27: 8475–8485.

17. Yao X, Balamurugan P, Arvey A, Leslie C, Zhang L. 2010. Heme controls the regulation of protein tyrosine kinases Jak2 and Src. *Biochem Biophys Res Commun* 403: 30–35.

18. Zhang L, Guarente L. 1995. Heme binds to a short sequence that serves a regulatory function in diverse proteins. *EMBO J* 14: 313–320.

19. Uma S, Matts RL, Guo Y, White S, Chen JJ. 2000. The N-terminal region of the heme-regulated eIF2alpha kinase is an autonomous heme binding domain. *Eur J Biochem* 267: 498–506.

20. Rafie-Kolpin M, Chefalo PJ, Hussain Z, Hahn J, Uma S, Matts RL, Chen JJ. 2000. Two heme-binding domains of heme-regulated eukaryotic initiation factor-2alpha kinase. N terminus and kinase insertion. *J Biol Chem* 275: 5171–5178.

21. Hach A, Hon T, Zhang L. 1999. A new class of repression modules is critical for heme regulation of the yeast transcriptional activator Hap1. *Mol Cell Biol* 19: 4324–4333.

22. Hira S, Tomita T, Matsui T, Igarashi K, Ikeda-Saito M. 2007. Bach1, a heme-dependent transcription factor, reveals presence of multiple heme binding sites with distinct coordination structure. *IUBMB Life* 59: 542–551.

23. Yang J, Ishimori K, O'Brian MR. 2005. Two heme binding sites are involved in the regulated degradation of the bacterial iron response regulator (Irr) protein. *J Biol Chem* 280: 7671–7676.

24. Yang J, Sangwan I, Lindemann A, Hauser F, Hennecke H, Fischer HM, O'Brian MR. 2006. Bradyrhizobium japonicum senses iron through the status of haem to regulate iron homeostasis and metabolism. *Mol Microbiol* 60: 427–437.

25. Hausott B, Kurnaz I, Gajovic S, Klimaschewski L. 2009. Signaling by neuronal tyrosine kinase receptors: Relevance for development and regeneration. *Anat Rec (Hoboken)* 292: 1976–1985.

26. Hartmann JT, Haap M, Kopp HG, Lipp HP. 2009. Tyrosine kinase inhibitors — a review on pharmacology, metabolism and side effects. *Curr Drug Metab* 10: 470–481.

27. Gentile A, Trusolino L, Comoglio PM. 2008. The Met tyrosine kinase receptor in development and cancer. *Cancer Metastasis Rev* 27: 85–94.

28. Quimby BB, Dasso M. 2003. The small GTPase Ran: Interpreting the signs. *Curr Opin Cell Biol* 15: 338–344.

29. Toksoz D, Merdek KD. 2002. The Rho small GTPase: Functions in health and disease. *Histol Histopathol* 17: 915–927.

30. Weissman B, Knudsen KE. 2009. Hijacking the chromatin remodeling machinery: Impact of SWI/SNF perturbations in cancer. *Cancer Res* 69: 8223–8230.

31. O'Malley BW, Kumar R. 2009. Nuclear receptor coregulators in cancer biology. *Cancer Res* 69: 8217–8222.

32. Spiegelman BM, Heinrich R. 2004. Biological control through regulated transcriptional coactivators. *Cell* 119: 157–167.

33. Castano J, Davalos V, Schwartz S, Jr., Arango D. 2008. EPH receptors in cancer. *Histol Histopathol* 23: 1011–1023.

34. Arvanitis D, Davy A. 2008. Eph/ephrin signaling: Networks. *Genes Dev* 22: 416–429.

35. Egea J, Klein R. 2007. Bidirectional Eph-ephrin signaling during axon guidance. *Trends Cell Biol* 17: 230–238.

36. Pasquale EB. 2008. Eph-ephrin bidirectional signaling in physiology and disease. *Cell* 133: 38–52.

37. Belhage B, Hansen GH, Elster L, Schousboe A. 1998. Effects of gamma-aminobutyric acid (GABA) on synaptogenesis and synaptic function. *Perspect Dev Neurobiol* 5: 235–246.

38. Ragan CI, McKernan RM, Wafford K, Whiting PJ. 1993. gamma-Aminobutyric acid-A (GABA-A) receptor/ion channel complex. *Biochem Soc Trans* 21 (Pt 3): 622–626.

39. Moult PR. 2009. Neuronal glutamate and GABAA receptor function in health and disease. *Biochem Soc Trans* 37: 1317–1322.

40. Anwyl R. 2009. Metabotropic glutamate receptor-dependent long-term potentiation. *Neuropharmacology* 56: 735–740.

41. Chernova T, Nicotera P, Smith AG. 2006. Heme deficiency is associated with senescence and causes suppression of N-methyl-D-aspartate receptor subunits expression in primary cortical neurons. *Mol Pharmacol* 69: 697–705.

42. Tang XD, Xu R, Reynolds MF, Garcia ML, Heinemann SH, Hoshi T. 2003. Haem can bind to and inhibit mammalian calcium-dependent Slo1 BK channels. *Nature* 425: 531–535.

43. Mendelsohn J. 2004. EGF receptors as a target for cancer therapy. *Trans Am Clin Climatol Assoc* 115: 249–253; discussion 53–54.

44. Arteaga CL. 2006. EGF receptor mutations in lung cancer: From humans to mice and maybe back to humans. *Cancer Cell* 9: 421–423.

45. Yamauchi M, Gotoh N. 2009. Theme: Oncology — molecular mechanisms determining the efficacy of EGF receptor-specific tyrosine kinase inhibitors help to identify biomarker candidates. *Biomark Med* 3: 139–151.

46. Reade CA, Ganti AK. 2009. EGFR targeted therapy in non-small cell lung cancer: Potential role of cetuximab. *Biologics* 3: 215–224.

47. Beltowski J, Lowicka E. 2009. EGF receptor as a drug target in arterial hypertension. *Mini Rev Med Chem* 9: 526–538.

48. Miettinen P, Ormio P, Hakonen E, Banerjee M, Otonkoski T. 2008. EGF receptor in pancreatic beta-cell mass regulation. *Biochem Soc Trans* 36: 280–285.

49. Sequist LV, Lynch TJ. 2008. EGFR tyrosine kinase inhibitors in lung cancer: An evolving story. *Annu Rev Med* 59: 429–442.

50. Badache A, Goncalves A. 2006. The ErbB2 signaling network as a target for breast cancer therapy. *J Mammary Gland Biol Neoplasia* 11: 13–25.

51. Melisko ME, Glantz M, Rugo HS. 2009. New challenges and opportunities in the management of brain metastases in patients with ErbB2-positive metastatic breast cancer. *Nat Clin Pract Oncol* 6: 25–33.

52. Sorkin A, Goh LK. 2009. Endocytosis and intracellular trafficking of ErbBs. *Exp Cell Res* 315: 683–696.

53. Yu D, Hung MC. 2000. Overexpression of ErbB2 in cancer and ErbB2-targeting strategies. *Oncogene* 19: 6115–6121.

54. Garratt AN, Ozcelik C, Birchmeier C. 2003. ErbB2 pathways in heart and neural diseases. *Trends Cardiovasc Med* 13: 80–86.

55. Baselga J, Swain SM. 2009. Novel anticancer targets: Revisiting ERBB2 and discovering ERBB3. *Nat Rev Cancer* 9: 463–475.

56. Baxter RC, Twigg SM. 2009. Actions of IGF binding proteins and related proteins in adipose tissue. *Trends Endocrinol Metab* 20: 499–505.

57. Gualco E, Wang JY, Del Valle L, Urbanska K, Peruzzi F, Khalili K, Amini S, Reiss K. 2009. IGF-IR in neuroprotection and brain tumors. *Front Biosci* 14: 352–375.

58. Tao Y, Pinzi V, Bourhis J, Deutsch E. 2007. Mechanisms of disease: Signaling of the insulin-like growth factor 1 receptor pathway — therapeutic perspectives in cancer. *Nat Clin Pract Oncol* 4: 591–602.

59. LeRoith D, Yakar S. 2007. Mechanisms of disease: Metabolic effects of growth hormone and insulin-like growth factor 1. *Nat Clin Pract Endocrinol Metab* 3: 302–310.

60. Freude S, Schilbach K, Schubert M. 2009. The role of IGF-1 receptor and insulin receptor signaling for the pathogenesis of Alzheimer's disease: From model organisms to human disease. *Curr Alzheimer Res* 6: 213–223.

61. Rowzee AM, Lazzarino DA, Rota L, Sun Z, Wood TL. 2008. IGF ligand and receptor regulation of mammary development. *J Mammary Gland Biol Neoplasia* 13: 361–370.

62. Abbas A, Grant PJ, Kearney MT. 2008. Role of IGF-1 in glucose regulation and cardiovascular disease. *Expert Rev Cardiovasc Ther* 6: 1135–1149.

63. Arai Y, Kojima T, Takayama M, Hirose N. 2009. The metabolic syndrome, IGF-1, and insulin action. *Mol Cell Endocrinol* 299: 124–128.

64. Werner H, Weinstein D, Bentov I. 2008. Similarities and differences between insulin and IGF-I: Structures, receptors, and signalling pathways. *Arch Physiol Biochem* 114: 17–22.

65. Gilfillan AM, Rivera J. 2009. The tyrosine kinase network regulating mast cell activation. *Immunol Rev* 228: 149–169.

66. Filippakopoulos P, Muller S, Knapp S. 2009. SH2 domains: Modulators of nonreceptor tyrosine kinase activity. *Curr Opin Struct Biol* 19: 643–649.

67. Mocsai A, Ruland J, Tybulewicz VL. 2010. The SYK tyrosine kinase: A crucial player in diverse biological functions. *Nat Rev Immunol* 10: 387–402.

68. Schlessinger J, Lemmon MA. 2003. SH2 and PTB domains in tyrosine kinase signaling. *Sci STKE* 2003: RE12.

69. Perona R. 2006. Cell signalling: Growth factors and tyrosine kinase receptors. *Clin Transl Oncol* 8: 77–82.

70. Csiszar A. 2006. Structural and functional diversity of adaptor proteins involved in tyrosine kinase signalling. *Bioessays* 28: 465–479.

71. Tohyama Y, Yamamura H. 2009. Protein tyrosine kinase, syk: A key player in phagocytic cells. *J Biochem* 145: 267–273.

72. Constantinescu SN, Girardot M, Pecquet C. 2008. Mining for JAK-STAT mutations in cancer. *Trends Biochem Sci* 33: 122–131.

73. Guarino M. 2010. Src signaling in cancer invasion. *J Cell Physiol* 223: 14–26.

74. Fizazi K. 2007. The role of Src in prostate cancer. *Ann Oncol* 18: 1765–1773.

75. Li S. 2007. Src kinase signaling in leukaemia. *Int J Biochem Cell Biol* 39: 1483–1488.

76. Verma A, Kambhampati S, Parmar S, Platanias LC. 2003. Jak family of kinases in cancer. *Cancer Metastasis Rev* 22: 423–434.

77. Lin J, Arlinghaus R. 2008. Activated c-Abl tyrosine kinase in malignant solid tumors. *Oncogene* 27: 4385–4391.

78. Yim EK, Siwko S, Lin SY. 2009. Exploring Rak tyrosine kinase function in breast cancer. *Cell Cycle* 8: 2360–2364.

79. Stirewalt DL, Meshinchi S. 2009. Receptor tyrosine kinase alterations in AML — biology and therapy. *Cancer Treat Res* 145: 85–108.

80. Valentino L, Pierre J. 2006. JAK/STAT signal transduction: Regulators and implication in hematological malignancies. *Biochem Pharmacol* 71: 713–721.

81. Wagh PK, Peace BE, Waltz SE. 2008. Met-related receptor tyrosine kinase Ron in tumor growth and metastasis. *Adv Cancer Res* 100: 1–33.

82. Bradshaw JM. 2010. The Src, Syk, and Tec family kinases: Distinct types of molecular switches. *Cell Signal* 22: 1175–1184.

83. Funakoshi-Tago M, Pelletier S, Moritake H, Parganas E, Ihle JN. 2008. Jak2 FERM domain interaction with the erythropoietin receptor regulates Jak2 kinase activity. *Mol Cell Biol* 28: 1792–1801.

84. Grote K, Luchtefeld M, Schieffer B. 2005. JANUS under stress — role of JAK/STAT signaling pathway in vascular diseases. *Vascul Pharmacol* 43: 357–363.

85. Haan C, Kreis S, Margue C, Behrmann I. 2006. Jaks and cytokine receptors — an intimate relationship. *Biochem Pharmacol* 72: 1538–1546.

86. Hall T, Emmons TL, Chrencik JE, Gormley JA, Weinberg RA, Leone JW, Hirsch JL, Saabye MJ, Schindler JF, Day JE, Williams JM, Kiefer JR, Lightle SA, Harris MS, Guru S, Fischer HD, Tomasselli AG. 2010. Expression, purification, characterization and crystallization of non- and phosphorylated states of JAK2 and JAK3 kinase domain. *Protein Expr Purif* 69: 54–63.

87. Igaz P, Toth S, Falus A. 2001. Biological and clinical significance of the JAK-STAT pathway; lessons from knockout mice. *Inflamm Res* 50: 435–441.

88. Kurdi M, Booz GW. 2009. JAK redux: A second look at the regulation and role of JAKs in the heart. *Am J Physiol Heart Circ Physiol* 297: H1545–H1556.

89. Rabkin R, Sun DF, Chen Y, Tan J, Schaefer F. 2005. Growth hormone resistance in uremia, a role for impaired JAK/STAT signaling. *Pediatr Nephrol* 20: 313–318.

90. Schindler CW. 2002. Series introduction. JAK-STAT signaling in human disease. *J Clin Invest* 109: 1133–1137.

91. Scott MJ, Godshall CJ, Cheadle WG. 2002. Jaks, STATs, Cytokines, and Sepsis. *Clin Diagn Lab Immunol* 9: 1153–1159.

92. Shuai K, Liu B. 2003. Regulation of JAK-STAT signaling in the immune system. *Nat Rev Immunol* 3: 900–911.

93. Ward AC, Touw I, Yoshimura A. 2000. The Jak-Stat pathway in normal and perturbed hematopoiesis. *Blood* 95: 19–29.

94. Wilks AF. 2008. The JAK kinases: not just another kinase drug discovery target. *Semin Cell Dev Biol* 19: 319–328.

95. Yamaoka K, Saharinen P, Pesu M, Holt VE, 3rd, Silvennoinen O, O'Shea JJ. 2004. The Janus kinases (Jaks). *Genome Biol* 5: 253.

96. Parganas E, Wang D, Stravopodis D, Topham DJ, Marine JC, Teglund S, Vanin EF, Bodner S, Colamonici OR, van Deursen JM, Grosveld G, Ihle JN. 1998. Jak2 is essential for signaling through a variety of cytokine receptors. *Cell* 93: 385–395.

97. Neubauer H, Cumano A, Muller M, Wu H, Huffstadt U, Pfeffer K. 1998. Jak2 deficiency defines an essential developmental checkpoint in definitive hematopoiesis. *Cell* 93: 397–409.

98. Pelletier S, Gingras S, Funakoshi-Tago M, Howell S, Ihle JN. 2006. Two domains of the erythropoietin receptor are sufficient for Jak2 binding/activation and function. *Mol Cell Biol* 26: 8527–8538.

99. Jamieson CH, Gotlib J, Durocher JA, Chao MP, Mariappan MR, Lay M, Jones C, Zehnder JL, Lilleberg SL, Weissman IL. 2006. The JAK2 V617F mutation occurs in hematopoietic stem cells in polycythemia vera and predisposes toward erythroid differentiation. *Proc Natl Acad Sci U S A* 103: 6224–6229.

100. Kroger N, Badbaran A, Holler E, Hahn J, Kobbe G, Bornhauser M, Reiter A, Zabelina T, Zander AR, Fehse B. 2007. Monitoring of the JAK2-V617F mutation by highly sensitive quantitative real-time PCR after allogeneic stem cell transplantation in patients with myelofibrosis. *Blood* 109: 1316–1321.

101. Li Z, Xing S, Wang S, Ho WT, Zhao ZJ. 2007. Characterization of a highly effective protein substrate for analysis of JAK2V617F activity. *Exp Hematol* 35: 1624–1632.

102. Li Z, Xu M, Xing S, Ho WT, Ishii T, Li Q, Fu X, Zhao ZJ. 2007. Erlotinib effectively inhibits JAK2V617F activity and polycythemia vera cell growth. *J Biol Chem* 282: 3428–3432.

103. Mossuz P, Arlotto M, Hermouet S, Bouamrani A, Lippert E, Girodon F, Dobo I, Vincent P, Cahn JY, Berger F. 2008. Proteomic study of the impact of the JAK2-V617F mutation on the phenotype of essential thrombocythemia. *Exp Hematol* 36: 1642–1647.

104. Olsen R, Dunphy C, O'Malley D, Rice L, Ewton A, Chang C-C. 2008. The implication of identifying JAK2 (V617F) in myeloproliferative neoplasms and myelodysplastic syndromes with bone marrow fibrosis. *J Hematopathol* 1: 111–117.

105. Williams DM, Kim AH, Rogers O, Spivak JL, Moliterno AR. 2007. Phenotypic variations and new mutations in JAK2 V617F-negative polycythemia vera, erythrocytosis, and idiopathic myelofibrosis. *Exp Hematol* 35: 1641–1646.

106. Xing S, Wanting TH, Zhao W, Ma J, Wang S, Xu X, Li Q, Fu X, Xu M, Zhao ZJ. 2008. Transgenic expression of JAK2V617F causes myeloproliferative disorders in mice. *Blood* 111: 5109–5117.

107. Xu X, Zhang Q, Luo J, Xing S, Li Q, Krantz SB, Fu X, Zhao ZJ. 2007. JAK2V617F: Prevalence in a large Chinese hospital population. *Blood* 109: 339–342.

108. Zhao R, Fu X, Teng L, Li Q, Zhao ZJ. 2003. Blocking the Function of Tyrosine Phosphatase SHP-2 by Targeting Its Src Homology 2 Domains. *J Biol Chem* 278: 42893–42898.

109. Zhao R, Xing S, Li Z, Fu X, Li Q, Krantz SB, Zhao ZJ. 2005. Identification of an Acquired JAK2 Mutation in Polycythemia Vera. *J Biol Chem* 280: 22788–22792.

110. Levine RL, Pardanani A, Tefferi A, Gilliland DG. 2007. Role of JAK2 in the pathogenesis and therapy of myeloproliferative disorders. *Nat Rev Cancer* 7: 673–683.

111. Anderson KE, Sassa S, Bishop DF, Desnick RJ. 2009. Disorders of heme biosynthesis: X-linked sideroblastic anemia and the porphyrias. In *The metabolic and molecular bases of inherited disease*, ed. CR Scriver, AL Beaudt, WS Sly, D Valle, C Barton, KW Kinzler, B Vogelstein, Chapter 124, pp. 1–53. New York: The McGraw-Hill Companies, Inc.

112. Araujo J, Logothetis C. 2009. Targeting Src signaling in metastatic bone disease. *Int J Cancer* 124: 1–6.

113. Brunton VG, Frame MC. 2008. Src and focal adhesion kinase as therapeutic targets in cancer. *Curr Opin Pharmacol* 8: 427–432.

114. Cooper JA, Qian H. 2008. A mechanism for SRC kinase-dependent signaling by noncatalytic receptors. *Biochemistry* 47: 5681–5688.

115. de Groot J, Milano V. 2009. Improving the prognosis for patients with glioblastoma: The rationale for targeting Src. *J Neurooncol* 95: 151–163.

116. Finn RS. 2008. Targeting Src in breast cancer. *Ann Oncol* 19: 1379–1386.

117. Giaccone G, Zucali PA. 2008. Src as a potential therapeutic target in non-small-cell lung cancer. *Ann Oncol* 19: 1219–1223.

118. Hu G, Minshall RD. 2009. Regulation of transendothelial permeability by Src kinase. *Microvasc Res* 77: 21–25.

119. Ingley E. 2008. Src family kinases: regulation of their activities, levels and identification of new pathways. *Biochim Biophys Acta* 1784: 56–65.

120. Lavoie JN, Landry MC, Faure RL, Champagne C. 2010. Src-family kinase signaling, actin-mediated membrane trafficking and organellar dynamics in the control of cell fate: Lessons to be learned from the adenovirus E4orf4 death factor. *Cell Signal* 22: 1604–1614.

121. Saad F. 2009. Src as a therapeutic target in men with prostate cancer and bone metastases. *BJU Int* 103: 434–440.

122. Salmond RJ, Filby A, Qureshi I, Caserta S, Zamoyska R. 2009. T-cell receptor proximal signaling via the Src-family kinases, Lck and Fyn, influences T-cell activation, differentiation, and tolerance. *Immunol Rev* 228: 9–22.

123. Sandilands E, Frame MC. 2008. Endosomal trafficking of Src tyrosine kinase. *Trends Cell Biol* 18: 322–329.

124. Seliger B, Massa C, Rini B, Ko J, Finke J. 2010. Antitumour and immune-adjuvant activities of protein-tyrosine kinase inhibitors. *Trends Mol Med* 16: 184–192.

Chapter **8**

THE CHEMICAL AND STRUCTURAL BASES OF HEME RECOGNITION

Binding Interactions of Heme with Proteins and Peptides

Y. Li and Li Zhang

Overview

Understanding heme recognition and its interactions with protein provides insights into how structures are related to heme biological functions. In this chapter, we will review the current structural information about hemeprotein binding interactions, highlighting the chemical and structural bases of heme recognition. Versatile heme-binding interactions, owing to the involvement of both the metal iron and the aromatic porphyrin ring, are the hallmark in heme recognition and play critical roles in both enzymatic and electron-transfer reactions. In addition, functionally important conformational changes modulated by heme-binding interactions are the underlining mechanisms of the heme functions in sensing and regulation. Reversible binding interactions and conformational changes are crucial to the triggering mechanism of loading and releasing small molecules or itself in the function of transport and storage. Here, we will summarize X-ray crystallographic studies that pertain to the structure–function relationships of heme proteins by revealing the chemical and structural bases of hemeprotein binding interactions.

Introduction

Heme (iron–protoporphyrin IX) and a family of related iron porphyrin molecules, shown in Fig. 1, bind to proteins through a combination of non-covalent interactions and up to two axial coordinative bonds to the iron by an amino acid side chain that contains oxygen, nitrogen or sulfur atoms. In the case of c-type heme, which is synthesized from heme b, the heme group is conjugated with the CX_nCH ($n = 2$–4) motif of protein through covalent thioether linkages between its vinyl groups and the cysteine side chain (1–2). Covalent heme binding is also present in CYP4 family of P450 enzymes and the mammalian peroxidases (3–4). Because of its remarkable bonding versatilities, heme can adapt to a wide variety of structural environments in a protein, which accounts for the diverse biological functions of heme groups. Some of these roles include the heme group activity in electron transfer, catalysis, binding of diatomic molecules (O_2, CO, NO sensing, transport and storage), ion-channel modulation (5–6), signal transduction and control of gene expression (7–9). There are also heme proteins dedicated to the storage and transport of heme molecules. Since the determination of the first myoglobin crystal structure (10), significant insights into the hemeprotein interactions have been gained through structural biology (11), site-directed mutagenesis (12–13), interrogation with ligand probes (14) and peptide–heme model systems (15–17). The structural environments of heme in globins and cytochrome enzymes have been investigated extensively and well-reviewed (18). There

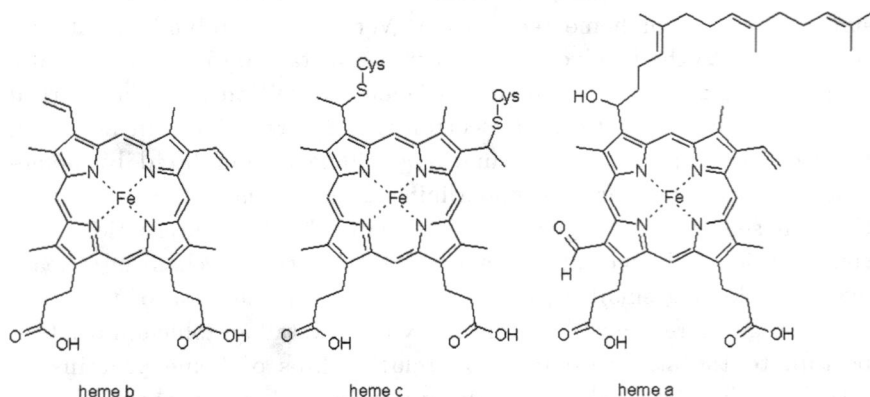

Fig. 1. The structures of ferric iron (Fe^{+3}) or ferrous iron (Fe^{+2}) heme b (protopophyrin IX), c-type and a-type heme.

are emerging details of a structural basis for understanding the mechanism of heme-mediated signal transduction (9, 12, 19–20).

A major approach to investigate hemeprotein interactions has long been studied by optical absorption spectra (21), resonance Raman spectra (13, 22), electron paramagnetic resonance (23–24) and nuclear magnetic resonance (25–26). Site-directed mutagenesis, together with biochemical and biophysical methods, provides a powerful tool to elucidate the precise roles of key residues in structure and function of hemeproteins (13, 21). For example, the replacement of a single histidine ligand residue by a cysteine or tyrosine would transform the heme oxygenase into an oxidase (27), illustrating the critical role of this proximal ligand in controlling reduction potential and ultimately the enzymatic function of the heme protein.

Another approach is to characterize the binding pockets that recognize and interact with a heme molecule by systematically examining all hemeprotein structures currently known. There are now over 2300 entries in the Protein Data Bank that contain a porphyrin ring, representing over 100 heme binding sites of non-redundant hemeproteins that remarkably feature over 30 distinct structural folds (28–29). These diverse scaffolds and architectures for heme-binding interactions constitute the structural basis of heme functions including modulation of redox potentials and the mechanism of regulatory hemeproteins. A database has been developed, linking electrochemical function of heme proteins to heme protein structural classifications including protein fold, heme types and heme axial ligands (30).

Additionally, progress made in the design of heme binding peptides and the *de novo* engineering of functional heme proteins with novel properties also contributed to our understanding of chemical and structural bases for heme recognition (15). Early spectrascopic investigation was reported on the formation of "red complex" between ferriheme and poly-L-lysine under basic conditions and complex formation between heme and polyhistidine and copoly-(histidine, glutamic acid) (31–32). Reversible interactions were observed between heme and a short synthetic peptide corresponding to the signature sequence of heme responsive motifs (HRM) present in some regulatory heme-binding proteins (33). For the HRM peptide interaction with heme, the cysteine residue in the peptide sequence was found to be essential, which presumably acts as a ligand to the heme iron. More recent studies of synthetic heme peptides (34–36) showed that heme binding with helix bundles designed with relatively simple rules revealed the intricacies of hemeprotein interactions. It is now possible to design and engineer protein-heme interactions of a heme-binding protein to switch from one biological function to another, for example, creating nitric oxide reductase from myoglobin (37) and an oxygen carrier from the mutants of a heme-binding protein human serum albumin (38–40) or artificial proteins (41–42).

Heme protein architectures

A comparative analysis of the representative b-type heme protein structures highlights a wide range of folds and architectures of heme-binding proteins (28–29). It also illustrates diverse structural environments of the heme pocket for heme recognition and interactions. Some representative heme-binding scaffolds are shown in Fig. 2. Although heme proteins adopt many distinct scaffolds to form heme-binding pockets, among the known

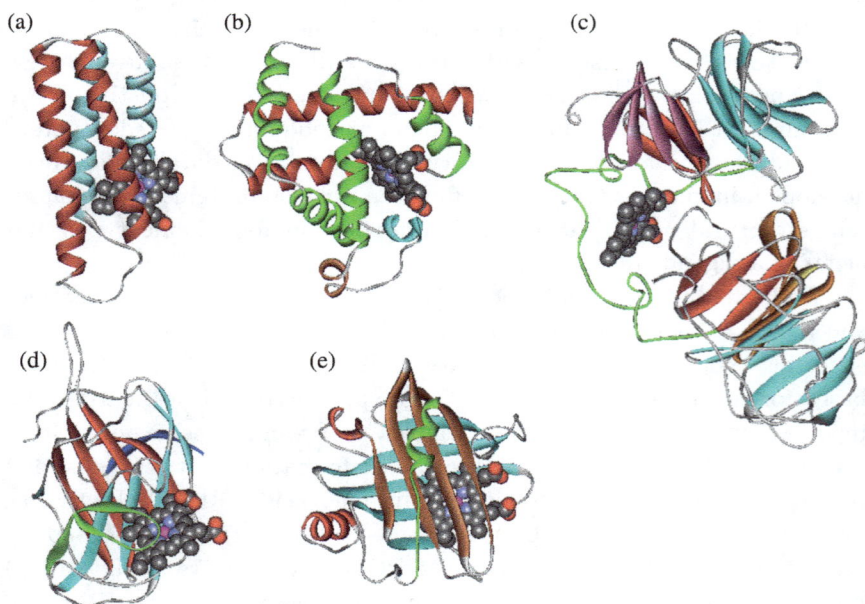

Fig. 2. Represetative scaffolds of heme binding proteins. (a) Up-down helical bundle of cytochrome b562 (PDB: 256b); (b) Orthogonal helical bundle of myoglobin (PDB: 1a6n); (c) β-propeller of hemopexin (PDB: 1qjs); (d) β-sandwich of cytochrome domain of cellobiose dehydrogenase (PDB: 1d7d); (e) β-barrel of nitrophorin 4 (PDB: 1d2u).

structures of b-type heme proteins, heme groups are often found in contact with an α-helical secondary structure parallel to the plane of the porphyrin ring or heme face. This is primarily due to the fact that an α-helix, with a cylindrical diameter comparable to the dimension of the porphyrin ring, provides strong packing interactions between protein and heme (18). For helical structures to interact with both faces of a heme group, α-helical bundles are commonly found in natural heme proteins and frequent template peptides designed for heme binding, usually with helical histidine residues forming the axial coordinative bond (43). The helices for heme binding interactions are bundled either up-down or orthogonal, as shown in Figs. 2a and 2b, respectively. Other heme environments include the β-propeller (Fig. 2c), β-sandwich (Fig. 2d), β-barrel (Fig. 2e), and loop regions or excursion from secondary elements (28–29).

One heme-binding architecture can be used for multiple functions, and conversely, different heme-binding architectures are also used by Nature for one particular type of biological function. For instance, heme oxygenase ChuS (44) and heme transport proteobacteria-protein HemS share the same protein architecture (45). The globin scaffold, found in hemoglobin and myoglobin for transporting and storing dioxygen, was also adopted in globin-coupled sensors for oxygen detection (46). For small molecule gas-sensor proteins, the heme-containing regulatory domains feature four different families of heme-binding modules: the heme-binding PAS (the Per-ARNT-Sim sequence motif) domain (47), globin-coupled sensor, CooA, and heme-NO-binding (HNOB) family proteins (48). These heme-binding modules, as shown in Fig. 3, confer exquisite sensitivity to gases (O_2, NO and CO) and mediate adaptive responses by coupling to a variety of neighboring transmitters that control either enzymatic activity, level of second-messenger molecules, or regulation of macromolecular interactions such as DNA binding. These heme-based sensors have been studied extensively and their characteristics and detailed mechanism of signal transduction have been extensively reviewed recently (48–49).

Heme exposure and solvent accessibility

In a survey of hemeprotein structures, the level of heme exposure, measured by its solvent accessible surface area (SASA), was found to vary significantly, from being highly solvent exposed (SASA > 250 Å² in comparison to 830 Å² of a free heme group) to almost completely buried

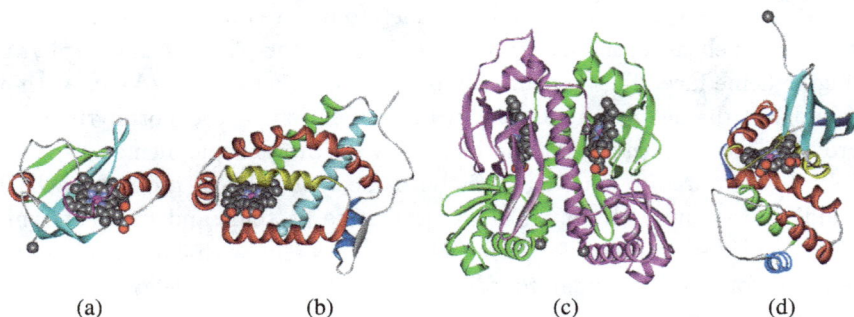

Fig. 3. Known scaffolds of small molecule gas-sensor proteins. (a) Heme-binding PAS domain of heme-based-sensor proteins FixL (PDB: 2vv6); (b) Globin-coupled heme-based sensor HemAT domain (PDB: 1or4); (c) CO-dependent transcription factor CooA (PDB: 2fmy shown as a dimer with the ribbon structures colored by monomer); (d) Heme-NO-binding domain related to soluble guanylate cyclases (PDB: 1u56).

(SASA around 2–3 $Å^2$) in a protein (28–29). Although the solvent accessibility of the heme group bound to a protein varies, it is characteristically related to heme protein functions and the role heme plays in the protein. Heme groups in proteins with a transient heme-binding function tend to have large solvent exposure, whereas heme groups in multiheme proteins and enzymes are often buried deeply. The heme group is solvent exposed intermediately (SASA ~100 $Å^2$) for proteins involved in electron transfer or diatomic molecule binding. This reflects different requirements of heme-binding sites for different function of heme proteins. When the bound heme group is exposed to solvent, it is usually the propionate groups that point toward the outside or surface of the protein.

Ligation of the heme iron

Ligation of the heme iron by the amino acid chain is the primary factor dictating the specific function of the heme protein. The heme iron in hemoproteins is coordinated by four nitrogen atoms of the tetrapyrrole rings of the porphrin, with additional coordination by one or two axial ligands from the amino acid side chain that contains oxygen, nitrogen or sulfur atoms, typically as histidine, cysteine, methionine, tyrosine or lysine. In rare cases, the axial ligand can be the N-terminus amino group of proline (50–51) or the amide group of asparagine (52). Considering that heme itself forms a dimer with iron–carboxylate coordination (53–54), iron-coordination by a carboxylate group rarely occurs in heme protein (30) with a few exceptions (37, 55–58).

When coordinated by only one axial ligand from the amino acid in proteins, the heme iron often binds to a small molecule, including molecular oxygen, nitric oxide and carbon monoxide. To sense or transport these small molecules, the fifth axial ligand is usually an imidazole from histidine, as found in hemoglobin and myoglogin. On the other hand, in cytochrome P450-dependent monooxygenases that involve the bound molecular oxygen in enzymatic reactions, the fifth axial ligand is usually the thiolate from a cysteine residue. Indeed, mutation of the histidine fifth axial ligand in hemoglobin and myoglobin to a cysteine causes a switch of the hemeprotein function from reversible oxygen binding to dioxygen activation (59). Unlike the heme-containing cytochrome P450-dependent monooxygenases, heme-containing dioxygenases are comprised of a proximal histidine (60–62). The fifth axial ligand is a histidine and a tyrosine residue in peroxidases and catalases, respectively, which are involved in hydrogen peroxide activation. For human heme oxygenase-1, for which heme is the substrate, replacement of the proximal histidine ligand by a cysteine or tyrosine, transformed it into an oxidase (27).

When the heme iron is bound to two axial amino-acid liagnds, heme proteins are usually involved in electron-transfer, as found in cytochromes c, characterized by a histidine/methionine axial ligand combination, and in cytochromes b, characterized by bis-histidine axial coordination. An atypical His-Tyr axial ligation was found in the diheme enzyme MauG (63).

Axial coordination of the heme iron by an amino acid side chain is believed to contribute in part, if not entirely, to the overall hemeprotein binding affinity. For example, an axial coordination to a histidine side chain would be comparable to the heme binding by an imidazole molecule, for which the binding energy was measured to be about 10 kcal/mol (64). However, site-directed mutagenesis studies of globin, for which the proximal histidine was replaced by a glycine, showed that the heme with its iron bound to an exogenous imidazole molecule still binds to the protein even though the heme iron is not coordinated by a protein side chain (65–66). Similarly, the H175G-mutant of cytochrome-c peroxidase still binds heme to which an exogenous imidazole (67–68) or a water molecule (69) is present in place of proximal histidine-175 in the wild type. In either mutant, the exogenous molecule occupying the cavity excavated by the removal of a proximal histidine side chain may also bring about stabilization of the hemeprotein complex by its non-covalent interactions with the protein (70–71).

A similar observation was made for heme oxygenase (72). Namely, replacement of the proximal histidine by an alanine did not cause a drastic structural change of the heme-bound protein complex, except the empty pocket created underneath the ferric ion in the heme and the loss of iron binding ligand (72). The alanine mutant lost 90% of its heme oxygenase activity, which can be restored by supplementing imidazole almost to the level of the wild type. Another example is the histidine-93 to a glycine mutant of sperm whale myoglobin. As a convenient system for studying ligand–protein interactions and the functional effect, the mutant H93G of sperm whale myoglobin provides an open cavity capable of accommodating various substituted imidazole proximal ligand and unusual oxygen donors such as benzolate or phenolate, which compared distinctively to myoglobin H93Y mutant (73–74).

The axial ligation of the heme iron can be modulated by the protonation state of the interacting ligand group. Protonation of the axial imidazole could cause dissociation of heme from the histidine ligands as demonstrated in the designed heme protein maquettes (75). Crystal structures of horse methemoglobin determined at two different pH levels showed that both proximal and distal histidine residues are the axial ligand at pH 5.4, whereas at pH 7.1 only the proximal histidine remains the axial ligand and the distal axial ligand is replaced by a water molecule (76). In the case of heme protein HasA, axial ligation by the tyrosine ligand was found to be pH dependent through the protonation states of a neighboring histidine residue (77–78). The molecular basis of Bohr effect, the observed relationship between pH and oxygen binding to hemoglobin during oxygen transport and storage, was attributed to the protonation states of histidine around the heme-binding pocket (79–82).

For a histidine ligand, the imidazole ring can adopt a wide range of orientations, which was found to have relatively even distribution for b-type hemes. However, due to geometric constraints, two preferred orientations were found for the histidine ligand of the CXXCH motif in cytochrome c proteins (43, 83–85). Moreover, two imdazole rings in bis-histidine hemes are often perpendicular rather than parallel to each other. Electrostatic interactions between the heme propionates and the ligands were suggested to play only a minor role in determining histidine heme ligand orientations. On the other hand, the orientation can be affected by the interaction of contact residues as observed in A67V mutation of cytochrome b5 (86). An analysis of non-redundant crystal structures of heme proteins with a helical histidine ligand revealed a strong sequence preference near the ligand histidine residue depending upon a particular histidine rotamer distribution. Consensus helical sequence templates were derived for each respective rotamer in order to provide optimal van der Waals interactions of side chain contact with the porphyrin ring (43).

Covalent linkage

The two vinyl groups on the porphyrin ring provide potential attachment points for heme to be covalently linked to proteins (1–2). Although the Cys-X-X-Cys-His (CXXCH) motif has recently been proposed to be a thiol/disulfide redox switch in the heme-binding domain of the human BK channel (87), the CXXCH motif is best known in cytochromes c for the two cysteine residues being covalently attached to two vinyl groups in heme, and the imidazole group of histidine acting as one of the ligands coordinating the iron (88–89). These covalently bound cytochromes c can be broadly classified into four major types (90). The largest group includes small soluble hemoproteins of a chain of about 100 amino acids with an axial ligand of a histidine near the N-terminus and a methionine axial ligand near the C-terminus. The second class of cytochromes c has a four-helix bundle structure with the heme attachment site near the C-terminus. The other classes include multiheme cytochromes and high molecular weight tetraheme reaction center cytochromes c, which contain the longer forms of the heme attachment motif (91–92) of CXXXCH (93) and CXXXXCH (94) sequence. There are other forms of covalent attachment for c-type heme, including covalent linkage to a CXXCK sequence with the lysine serving as an axial ligand to the heme iron, or covalent linkage to a single cysteine of AXXCH or FXXCH motif by a single thioether bond (95).

The CXXCH motif, when present in α-helical structures, positions the two cysteine residues on the same side of the helix just one helix turn apart. These two cysteine residues, as shown in Fig. 4a, are geometrically ideal to form a macrocycle to the heme group by enzyme-catalyzed (89, 96) formation of two thioether bonds. The helix structure often comes to an end and turns into a loop conformation at the second cysteine so that the histidine can orient itself to make an axial ligand, as shown in Fig. 4b. For (A/F)XXCH motif, which forms a single thioether bond with the heme, it

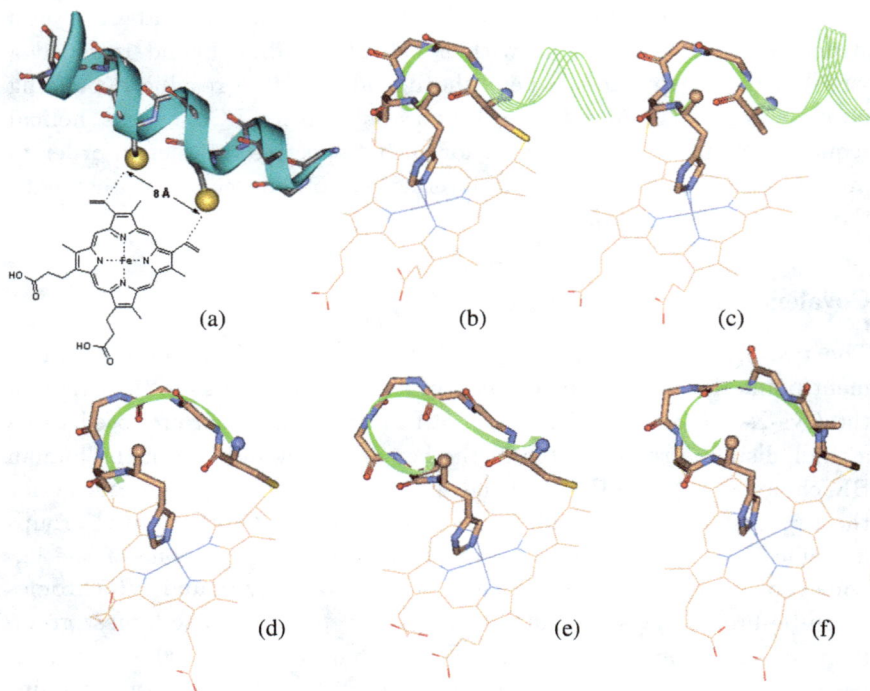

Fig. 4. Covalent conjugation of the cysteine residues with heme c. (a) macrocyclic cyclization of heme with a helical -CXXC-; (b) The crystal structures of heme proteins with CXXCH (PDB: 3m97) and (c) AXXCH motifs (PDB: 2w9k); (d) The crystal structures of heme proteins with CXXXCH (PDB: 1gyo), and (e–f) CXXXXCH motifs (PDB: 1gmb).

adopts similar conformation as that of CXXCH except the missing thioether linkage between the first cysteine and the heme, as shown in Fig. 4c. For the longer forms of the heme attachment motifs, they adopt turn structures for the backbone of the two cysteine residues and those in between, as shown in Figs. 4d–f.

   The mammalian peroxidases myeloperoxidase, ecosinophil peroxidase, lactoperoxidase and thyroid peroxidase represent a different class of covalently bound hemoproteins. Heme is covalently attached to these proteins through two covalent ester bonds (4) with an additional vinyl-sulfonium bond uniquely found in myeloperoxidase (97). Similarly, a covalent ester bond linking heme to the protein exists in the CYP4 family of P450 enzymes (3).

Hydrophobic and electrostatic interactions

Non-covalent interactions between heme and proteins involve both hydrophobic and electrostatic interactions. The methyl and vinyl groups along with the tetrapyrrole of the porphyrin ring are hydrophobic. The propionate groups on the porphyrin ring often form hydrogen bonds or salt bridges with amino acid residues in enzymes or with solvent molecules.

Hydrophobic interactions were surveyed by the contacts of the b-type heme with various non-polar residues (12). Leucine and isoleucine, along with valine, are the most common amino acids in close contact with the "heme face" of either proximal or distal sides. Distal phenylalanine, tryrosine and tryptophan are also commonly found in contact with the heme face, mainly by aromatic-aromatic π-stacking interactions with the porphyrin ring. When a tryptophan or phenylalanine is next to the proximal histidine, however, it adopts face-to-face conformation with the histidine imidazole ring, and packs edge-to-face to the porphyrin as in the crystal structures of hemopexin and cytochrome c peroxidase. It was speculated that the aromatic-imidazole π-stacking may help to orientate the proximal histidine sidechain to maintain bonding interactions of the histidine-iron bond.

Surrounding the heme edge of the porphyrin, often lies a phenylalanine or leucine side chain that makes van der Waals contacts (12). Alanine and valine are the other amino acids also frequently found in close contact with the non-polar sides of the heme. In contrast to its involvement in heme face interactions, isoleucine is rarely observed in contact with the heme edge. It was suggested that perhaps an isoleucine side chain is conformationally too flexible to lock the heme in a fixed position.

Binding interactions between α-helical peptides and the porphyrin face were found to correlate with hydrophobicity of the designed peptides with multiple alanine residues that are prone to form α-helices. Several linear peptides including a histidine as the ligating residue were found to form 2:1 complexes with coproporphyrin-I-atoiron(III), a soluble heme analog less likely to aggregate. The porphyrin ring is presumably situated in a hydrophobic pocket sandwiched by two α-helices of the designed peptide (34–35). Equilibrium binding constants measured in aqueous solutions were found to increase as a function of the number of hydrophobic residues, taken as the sum of alanine and phenylalanine, present in the designed peptides.

The heme propionates play an active role in regulating electron distribution to the iron center as well as biochemical properties of the heme

(19, 98). They are involved universally in electrostatic interactions, mainly through salt-bridges with arginine residues of the protein. Even for those bound heme groups that are highly solvent exposed, the propionate groups were found to interact with arginines (29). Other partners found to interact with the negatively charged propionates include lysine, histidine as well as tyrosine residues. These electrostatic interactions provide an anchoring mechanism for the orientation of heme binding to the protein by directing propionate of heme either towards the protein interior when deeply buried or towards outside when exposed (29).

Reversible heme-binding protein

Unlike c-type heme, which is covalently bound to hemoproteins, a- and b-type heme groups are not permanently attached to the protein. They are more readily dissociated reversibly from hemoproteins, which are referred as those that show transient heme binding. This reversible binding of heme to proteins is a fundamental property in heme cellular signal transduction, heme transportation and storage (9). Although heme is essential to life, excessive free intracellular heme is highly toxic because of its role in free radical generation. Therefore, the concentration of intracellular heme needs to be tightly controlled at low levels. There are multiple cellular mechanisms to regulate intracellular heme levels. First, heme downregulates its own synthesis by binding to 5-aminolevulinate synthase (ALAS), which is an enzyme catalyzing the rate-limiting reaction in heme synthesis (99–101). The crystal structure of *Rhodobacter capsulatus* ALAS has been reported (102–103), which corresponds to a truncated human ALAS with the heme-responsive motif sequence missing. Heme down regulates its synthesis by binding to the heme-responsive motif of the enzyme that prevents its translocation into mitochondria. Another negative feedback mechanism was proposed recently through heme binding to the nuclear receptor REV-ERBα to repress the transcription of the coactivator PGC-1α, a potent inducer of heme synthesis (104). Secondly, heme binds to the transcription factor Bach1 and induces the expression of heme oxygenase 1, which catalyzes heme degradation. To protect against dangerously high levels of free heme in humans, there are serum hemoproteins to sequester heme. Thus, the majority of heme in humans exist as protein bound by either hemopexin or human serum albumin (HSA), which is the preferred stable heme preparation used in the treatment of acute intermittent porphyria (105).

Heme-bound crystal structures of both human serum albumin (106–107) and hemopexin (108) are now available. The crystal structure of human serum albumin complex with heme showed the heme bound to a single site of the completely hydrophobic cavity (106–107). Investigation of heme-HSA unfolding by proton NMR and spectroscopic probe indicated a reversible, two-step unfolding of heme–human serum albumin (109). In contrast to the HSA heme-binding pockets, which involves mainly hydrophobic interactions, non-bonded hydrophobic interactions between heme and hemopexin are among the fewest in comparison to those observed in the other hemoproteins. Heme binds HSA with a low affinity, but binds hemopexin with a high affinity (29). The heme-binding pocket of hemopexin is composed of two β-propeller domains and the loop connecting them as shown in Fig. 2c. The heme group is highly solvent exposed but forms a significant number of hydrogen bonds with one of the propionate groups buried as shown in Fig. 5 by LIGPLOT (110). This characteristic of hydrogen bonds, rather than hydrophobic interactions, provided a transient heme-binding mechanism that could be easily modulated by pH. Changes in pH affect hydrogen bonds and heme is released from heme–hemopexin complex at low pH (111). Furthermore, spectroscopic studies suggested that multiple binding modes are involved in ferri- and ferro- and CO-ferro heme–hemopexin complexes (112–113).

A two-domain protein structure has been proposed as a universal model to a successful transient heme-binding system, which includes a flexible hinge connecting the two domains. The flexible hinge allows these domains to clamp onto heme with suitable binding interactions and an appropriate trigger for release (114). Hemopexin illustrates such a system in which its association with heme is found to be pH dependent. Hemopexin binds heme with a high affinity (Kd < 1 pM) but releases it under acidic condition when the pH is below 5, due to the disruption of salt-bridges and polar interactions between protein residues and the propionate groups (108, 111, 115).

Available data seem to support such a hinge linked, two-domain structure for other transient heme-binding proteins as well. By comparing apo structures with the corresponding heme-bound structures, differences were found between transient binding proteins and those that bind heme more or less permanently (29). Proteins of disordered apo structures with molten globule-like characteristics, which achieve significant structure ordering upon heme interactions, tend to bind heme permanently. On the other hand, transient heme-binding proteins tend to be fully folded in both apo

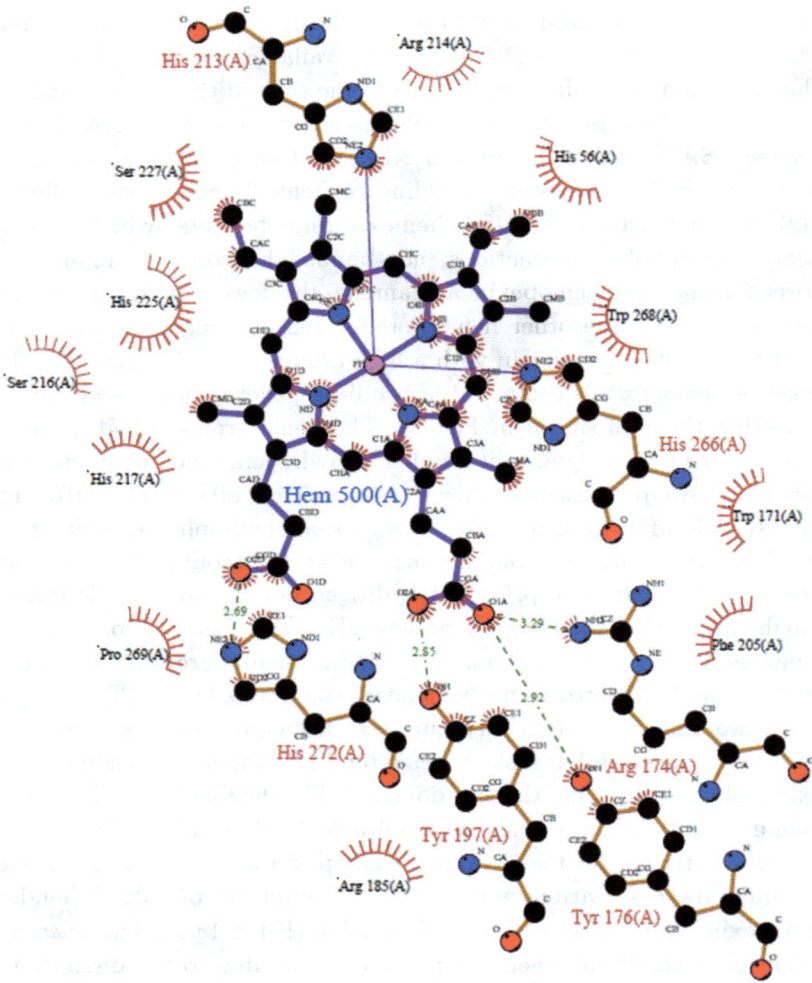

Fig. 5. LIGPLOT of the hydrophilic pocket for heme binding in hemopexin (PDB: 1qjs).

and holo states. For transient heme-binding proteins, the only change that occurs upon heme binding is the relative orientation of folded units through a global rigid body-like movement of the folded domains. For example, the bacteria heme-binding protein HasA is a reversible heme-binding protein, which captures a ferric b-type heme from the host's hemoglobin or free heme with high affinity (Kd ~ 10 pM). The crystal structure

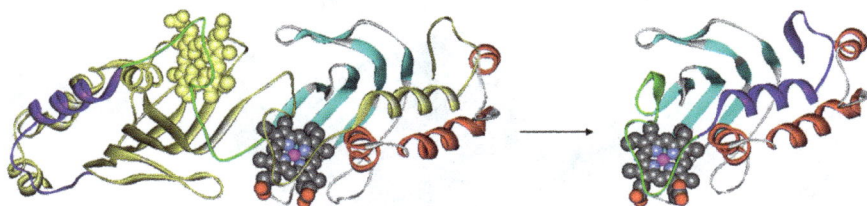

Fig. 6. Crystal structures of dimeric HasA in an "open" state (left with one monomer colored yellow, PDB: 2cn4) by domain swapping of the green loop and purple helix to the closed state (right, PDB: 1b2v).

of domain-swapped dimeric HasA-heme complex and crystal structure of monomeric HasA-heme complex showed, respectively, the open and closed forms of HasA protein. It provides a mechanistic model of how HasA protein captures the bound heme by the movement of one loop connecting well-defined units from an open to the closed state, as shown in Fig. 6 (116–117). Similarly, heme binding of the heme transport protein HemS involves a closing swing of the folded N- and C-terminal domains, as revealed by the crystal structures of apo and heme-bound HemS shown in Fig. 7 (45).

Structure and reduction potentials

The heme iron can adopt multiple oxidation states, the +2 reduced ferrous and +3 oxidized ferric states, and the less stable +4 state in certain catalytic intermediates. There are two electron spin moments for the ferric porphyrin complexes, that of one unpaired electron low spin or five unpaired electron high spin states (118–119). There is a rare third 3/2 spin state of three unpaired electrons suggested for the ferric heme. For the ferrous heme, there are also low (all paired electron) and high (four unpaired electrons) spin states, which can be characterized by optical spectra, magnetic susceptibility and oxidation–reduction titration (120–121).

Reduction potential values are dictated by differences in free energy between the oxidized ferric and reduced ferrous heme. By preferentially binding to ferric Fe(III) heme, hemoprotein would generally possess more negative reduction potentials, whereas preferential binding to ferrous Fe(II) heme would shift reduction potential to the positive end. Thus, the hemeprotein bonding interactions can modulate the reduction potentials

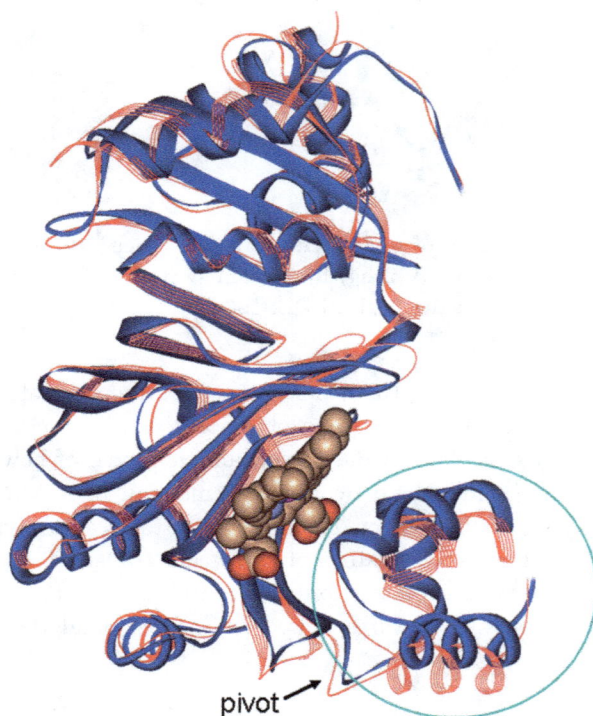

Fig. 7. Domain movement revealed by the crystal structures of apo HemS (red line ribbons, PDB: 2j0r) and its complex with heme (protein in blue ribbon, heme as space-filling model, PDB: 2j0p).

and thus enable heme enzymes to perform electron transfer in oxidation or reductive reactions.

The redox potentials of heme proteins are influenced by the nature of axial ligand (27, 122), solvent accessibility of the heme group (123–124), out-of-plane bending and distortion of the heme group (125–127), the local protein environments such as electrostatic potential and the protonation states of the heme propionate groups (128–130). Hemeprotein reduction potentials span a wide range of −550 mV to +450 mV versus the standard hydrogen electrode. For a given pair of axial ligands, the hemeprotein reduction potentials can vary widely but it follows the general trend from most positive to negative: His/Met > His/OH > Met/Met > Lys/(−) > His/His > Cys/(−) > Tyr/(−) > His/Tyr (30). This trend is consistent with the prediction of Hard-Soft Acid-Base theory which claims that hard bases preferentially stabilize ferric heme, resulting in negative shifts in reduction potentials. Ligation by a thiolate or phenolate results in a negative reduction potential. Soft ligands such as methionine preferentially stabilize ferrous heme and thus shift the redox potential to be more positive. For the same axial ligand, the reduction potentials can vary greatly too. For example, the reduction potentials of heme protein with histidine axial ligand(s) cover almost the entire range from -400 mV to $+400$ mV, presumably controlled by other factors such as the pH level (protonation state), solvent accessibility and electrostatic potentials. It is ascribed that positively charged residues around the heme in hemoglobin and myoglobin favor a low oxidation state of the iron for it to bind dioxygen. Similarly, a gradual increase in the ferric–ferrous redox potential may stem from the progressive change from a negatively charged surrounding of the heme-binding pocket in cytochromes b to a positive one in cytochromes c. More quantitatively, a linear relationship was observed between the anionic character of axial histidine and the midpoint reduction potential in c-type cytochromes (131).

In terms of how iron oxidation states affect the heme binding pockets, several crystallographic studies showed subtle conformational changes between different oxidation states (132–137). The conformational changes are sometime due a displacement of the axial ligand by either an exogenous ligand (138–139) or a solvent water molecule (140–141).

Heme-mediated signal transduction

The role heme plays in the regulation of gene expression and cellular signaling involves either direct heme-mediated signal transduction by reversible binding of heme to proteins, or heme-based sensor proteins in adaptive responses to the fluctuation of oxygen, carbon monoxide and nitric oxide levels. For example, binding of heme to Hap1, a yeast transcriptional regulator protein, causes Hap1 to be released from a high molecular complex (HMC) to form a Hap1 dimer, which binds DNA to activate transcription of several genes involved in cellular respiration and oxidative damage control. Reversal of heme binding or dissociation of heme from Hap1 allows HMC formation and terminates the transcriptional activity. In mammals, heme binds directly to the transcription factor BACH1, which controls the expression of heme oxygenase 1. These transcription factors contain potential heme-binding sites with the signature cysteine-proline (CP) heme responsive motifs. A short synthetic peptide based on these motifs was found to bind reversibly with heme (33). A similar heme regulatory motif is present in the mammalian iron regulatory protein 2 (IRP2), but conflicting results were reported about the HRM's role in heme-induced degradation of IRP2 (142–143). Spectroscopic studies of truncated peptides, containing the domain required for the heme-dependent degradation, suggest that a redox-dependent axial ligand exchange between the cysteine and histidine residues, the binding of cysteine to ferric heme and the binding of histidine to ferrous, respectively, is critical for heme-mediated IRP2 degradation (142). A more recent study, however, showed that the binding between heme and the intact IRP2 involves a non-specific interaction, which is unlikely to provide an efficient heme sensor *in vivo* (143). Since atomic structures are currently unavailable for all these proteins, it waits further investigation to reveal how heme interacts with these proteins at the binding site and the mechanism by which the heme-mediated signal transduction occurs.

On the other hand, there have been significant structural studies of heme-based sensor proteins, revealing detailed information about the signal transduction mechanism of these signal transducers. Heme-based sensor proteins contain a regulatory heme-binding module, coupled with a neighboring responding activation domain such as histidine protein kinases, cyclic nucleotide phosphodiesterase, chemotaxis methyl-carrier protein receptors, and transcription factors of the basic helix-loop-helix and helix-turn-helix classes. Four different types of heme-binding modules are currently known, including the heme PAS domain, globin-coupled sensor, CooA, and heme-NO-binding (HNOB) family proteins.

Crystal structures of the heme PAS domains are available for heme-based sensor proteins FixL (144–148) and EcDos (148–149). FixL is an oxygen sensing histidine kinase that regulates the expression of the genes related to nitrogen fixation. Phosphodiesterase (EcDos) from *Escherichia coli* is a gas-sensor enzyme in which binding of gas molecules, such as O_2, CO, and NO, to the Fe(II)-protoporphyrin IX complex in the sensor domain stimulates phosphodiesterase activity toward cyclic di-GMP (148–153). The mammalian transcription factor NPAS2, in response to CO levels, is another example of heme-based sensor protein containing the heme PAS domains. The heme irons of these heme proteins containing the PAS domain are coordinated by a proximal histidine, the most conserved residue of heme-PAS proteins. The heme iron is also coordinated by a displaceable methionine residue on the distal side in EcDos. This distal displaceable residue has been proposed as the main regulatory trigger, although a more recent model suggests that the entry of arginine into the heme distal pocket also plays a role as a trigger, by forming hydrogen bonds to the bound oxygen ligand (154). It has also been suggested that the monomer–dimer equilibrium controlled by the heme PAS domains may play a role in the signaling mechanism, since these proteins form dimer with a contact surface residing in the PAS domain itself (147). A general model involving helix-swap was also proposed as a triggering mechanism for signal transduction by heme PAS domains (48).

HemAT proteins exemplify the globin-coupled heme-based sensors that include more than 30 members. Although crystal structures of HemAT-Bs are known (155–156), it is unclear how subtle conformational changes are related to the switching or signaling of the proteins. There are indications that two distinct ligand-binding sites exist and a dynamic allosteric coupling between these sites is a trigger of a conformational switch for signaling in the O2-sensing protein HemAT (157).

The CooA protein is a CO-sensing transcriptional factor that regulates the oxidation of carbon monoxide in *Rhodospirillum rubrum*. Extensive structural and biochemical investigations have been carried out to elucidate the molecular mechanism of CO dependent activation (51, 158–163). The crystal structure of CooA showed a homodimer with the His 77 residue as the proximal axial ligand and the N-terminal amino group from each partner monomer providing the distal axial ligand. The CO or an imidazole binding to replace the distal axial N-terminal amino ligand resulted in the release of the N-terminus (Fig. 2c), which was found to be necessary but insufficient to achieve activation (162).

In response to nitric oxide and carbon monoxide, the soluble guanylate cyclase (sGC) converts GTP to the second messager cGMP, which exerts effects in many downstream processes (164–165). The sGC enzyme contains four domains, including the H-NOX sensor domain and the catalytic domain, connected by a coiled–coil domain (166). There are reported crystal structures of both catalytic (167) and coiled–coil domains (168). The structure of H-NOX domain is still unavailable for sGC, but there are crystal structures reported for several sGC homologs (169–171). Many models have been proposed to provide insights into the molecular basis of NO dependent activation (172). Based upon the crystal structure of cytochrom c, preferential binding to distal face by CO or displacement of the proximal histidine by NO was proposed as a mechanistic model for the heme-binding domain to explain the differential effects of NO and CO on the sGC enzymatic activity (173). Multistep mechanism of sGC activation was proposed, involving multiple NO-binding sites in light of the crystal structure of prokaryotic sGC homolog (174). More recently, crystal structures of apo, NO-bound and CO-bound H-NOX domains of a cyanobacterial homolog, suggested a molecular ruler mechanism with differential heme pivoting and heme bending (175). Distortion of the bound heme cofactor was suggested as a mechanism for signal transduction (176). Various computational simulations and energetic analyses were applied to model activation processes (177–178).

There are other heme-based proteins that transduce signaling upon sensing small gas molecules. The activity of heme protein cystathionine β-synthase is modulated by ligand controlled redox change and in response to carbon monoxide binding (179–181). The crystal structure of DNR (dissimilative nitrate respiration regulator), a putative NO-sensing bacterial transcriptional regulator, revealed a large hydrophobic cavity that may bind heme, yielding a heme-containing protein (182). A heme-binding protein is likely involved in the molecular origin of oxygen or carbon monoxide sensitivities of voltage-gated potassium channels (5–6, 9, 183–185).

Heme was identified as a ligand for the orphan nuclear hormone receptors REV-ERBα and REV-ERBβ (186–187), which are known to be critical components in the circadian rhythm. Heme binds reversibly to REV-ERB, leading to the recruitment of the N-CoR corepressor to repress expression of target genes including ARNTL involved in the circadian oscillator (188–189). The nuclear hormone receptors, REV-ERB, upon heme binding, suppress hepatic gluconeogenic gene expression and glucose output, thus playing an important role in adipogenesis (187, 190). The crystal structure of the heme-binding domain of REV-ERB has been reported (20, 191–192). Although heme binding with REV-ERB causes repression, the co-crystal structure of REV-ERBα ligand-binding domain with the nuclear receptor corepressor (NCOR) interaction domain suggested that REV-ERBα is a constitutive repressor that could mediate repression in the absences of heme (191–192). More recent studies showed that transcription repression mediated by heme-bound REV-ERB can be reversed by adding nitric oxide, which causes redox-dependent ligand switching and NO-induced ligand displacement (20, 193). In the heme-bound REV-ERB, the heme iron is coordinated by a cysteine and a histidine as the axial ligands in the oxidized ferric heme form. When reduced, the cysteine residue is no longer coordinating the ferrous heme iron, which is open for binding to either NO or CO. This suggested that cellular function of REV-ERB can be under heme, redox, or gas-regulated control.

There has been tremendous progress in the structure biology of heme proteins that advanced our understanding of structure–function relationships for heme proteins. The chemical and structural bases of hemeprotein binding interactions are now known to be critical to the heme functions involving enzymatic and electron transfer reactions. There are emerging views that conformational changes modulated by heme binding are fundamental to understand the mechanism and structure–function relationships of heme regulatory proteins. Extensive functional and structural studies, together with the discovery of new hemeproteins, continue to further our understanding of these diverse and biologically important molecules.

References

1. Barker PD, Ferguson SJ. 1999. Still a puzzle: Why is haem covalently attached in c-type cytochromes? *Structure* 7: R281–290.

2. Bowman SE, Bren KL. 2008. The chemistry and biochemistry of heme c: Functional bases for covalent attachment. *Nat Prod Rep* 25: 1118–1130.

3. Ortiz de Montellano PR. 2008. Mechanism and role of covalent heme binding in the CYP4 family of P450 enzymes and the mammalian peroxidases. *Drug Metab Rev* 40: 405–426.

4. Zederbauer M, Furtmuller PG, Brogioni S, Jakopitsch C, Smulevich G, Obinger C. 2007. Heme to protein linkages in mammalian peroxidases: Impact on spectroscopic, redox and catalytic properties. *Nat Prod Rep* 24: 571–584.

5. Tang XD, Xu R, Reynolds MF, Garcia ML, Heinemann SH, Hoshi T. 2003. Haem can bind to and inhibit mammalian calcium-dependent Slo1 BK channels. *Nature* 425: 531–535.

6. Hou S, Xu R, Heinemann SH, Hoshi T. 2008. The RCK1 high-affinity Ca2+ sensor confers carbon monoxide sensitivity to Slo1 BK channels. *Proc Natl Acad Sci U S A* 105: 4039–4043.

7. Zhang L, Hach A. 1999. Molecular mechanism of heme signaling in yeast: The transcriptional activator Hap1 serves as the key mediator. *Cell Mol Life Sci* 56: 415–426.

8. Mense SM, Zhang L. 2006. Heme: A versatile signaling molecule controlling The activities of diverse regulators ranging from transcription factors to MAP kinases. *Cell Res* 16: 681–692.

9. Hou S, Reynolds MF, Horrigan FT, Heinemann SH, Hoshi T. 2006. Reversible binding of heme to proteins in cellular signal transduction. *Acc Chem Res* 39: 918–924.

10. Kendrew JC, Bodo G, Dintzis HM, Parrish RG, Wyckoff H, Phillips DC. 1958. A three-dimensional model of the myoglobin molecule obtained by X-ray analysis. *Nature* 181: 662–666.

11. Poulos TL, Li H, Raman CS. 1999. Heme-mediated oxygen activation in biology: Cytochrome c oxidase and nitric oxide synthase. *Curr Opin Chem Biol* 3: 131–137.

12. Paoli M, Marles-Wright J, Smith A. 2002. Structure-function relationships in heme-proteins. *DNA Cell Biol* 21: 271–280.

13. Smulevich G, Feis A, Howes BD. 2005. Fifteen years of Raman spectroscopy of engineered heme containing peroxidases: What have we learned? *Acc Chem Res* 38: 433–440.

14. Anderson JL, Chapman SK. 2005. Ligand probes for heme proteins. *Dalton Trans*: 13–24.

15. Lombardi A, Nastri F, Pavone V. 2001. Peptide-based heme-protein models. *Chem Rev* 101: 3165–3189.

16. Collman JP. 1997. Functional analogs of heme protein active sites. *Inorganic Chemistry* 36: 5145–5155.

17. Collman JP, Boulatov R, Sunderland CJ, Fu L. 2004. Functional analogues of cytochrome c oxidase, myoglobin, and hemoglobin. *Chem Rev* 104: 561–588.

18. Landfried DA, Vuletich DA, Pond MP, Lecomte JT. 2007. Structural and thermodynamic consequences of b heme binding for monomeric apoglobins and other apoproteins. *Gene* 398: 12–28.

19. Poulos TL. 2007. The Janus nature of heme. *Nat Prod Rep* 24: 504–510.

20. Pardee KI, Xu X, Reinking J, Schuetz A, Dong A, Liu S, Zhang R, Tiefenbach J, Lajoie G, Plotnikov AN, Botchkarev A, Krause HM, Edwards A. 2009. The structural basis of gas-responsive transcription by the human nuclear hormone receptor REV-ERBbeta. *PLoS Biol* 7: e43.

21. Nienhaus K, Nienhaus GU. 2005. Probing heme protein-ligand interactions by UV/visible absorption spectroscopy. *Methods Mol Biol* 305: 215–242.

22. Spiro TG. 1985. Resonance Raman spectroscopy as a probe of heme protein structure and dynamics. *Adv Protein Chem* 37: 111–159.

23. Blumberg WE, Peisach J, Wittenberg BA, Wittenberg JB. 1968. The electronic structure of protoheme proteins. I. An electron paramagnetic resonance and optical study of horseradish peroxidase and its derivatives. *J Biol Chem* 243: 1854–1862.

24. Wittenberg BA, Kampa L, Wittenberg JB, Blumberg WE, Peisach J. 1968. The electronic structure of protoheme proteins. II. An electron paramagnetic resonance and optical study of cytochrome c peroxidase and its derivatives. *J Biol Chem* 243: 1863–1870.

25. Simonneaux G, Bondon A. 2005. Mechanism of electron transfer in heme proteins and models: The NMR approach. *Chem Rev* 105: 2627–2646.

26. Roberts GC. 1996. The other kind of biological NMR — studies of enzyme–substrate interactions. *Neurochem Res* 21: 1117–1124.

27. Liu Y, Moenne-Loccoz P, Hildebrand DP, Wilks A, Loehr TM, Mauk AG, Ortiz de Montellano PR. 1999. Replacement of the proximal histidine iron ligand by a cysteine or tyrosine converts heme oxygenase to an oxidase. *Biochemistry* 38: 3733–3743.

28. Schneider S, Marles-Wright J, Sharp KH, Paoli M. 2007. Diversity and conservation of interactions for binding heme in b-type heme proteins. *Nat Prod Rep* 24: 621–630.

29. Smith LJ, Kahraman A, Thornton JM. 2010. Heme proteins — diversity in structural characteristics, function, and folding. *Proteins* 78: 2349–2368.

30. Reedy CJ, Elvekrog MM, Gibney BR. 2008. Development of a heme protein structure-electrochemical function database. *Nucleic Acids Res* 36: D307–313.

31. Tohjo M, Shibata K. 1963. Absorption spectra of hematin complexes with polyhistidine and copoly-(histidine, glutamic acid). *Arch Biochem Biophys* 103: 401–408.

32. Blauer G. 1964. Spectrophotometric Investigations of the system poly-L-lysine: Ferriheme at pH 10–12. *Biochim Biophys Acta* 79: 547–562.

33. Zhang L, Guarente L. 1995. Heme binds to a short sequence that serves a regulatory function in diverse proteins. *EMBO J* 14: 313–320.

34. Huffman DL, Rosenblatt MM, Suslick KS. 1998. Synthetic heme–peptide complexes. *J Am Chem Soc* 120: 6183–6184.

35. Huffman DL, Suslick KS. 2000. Hydrophobic interactions in metalloporphyrin–peptide complexes. *Inorganic Chemistry* 39: 5418–5419.

36. Cowley AB, Kennedy ML, Silchenko S, Lukat-Rodgers GS, Rodgers KR, Benson DR. 2006. Insight into heme protein redox potential control and functional aspects of six-coordinate ligand-sensing heme proteins from studies of synthetic heme peptides. *Inorganic Chemistry* 45: 9985–10001.

37. Lin YW, Yeung N, Gao YG, Miner KD, Tian S, Robinson H, Lu Y. 2010. Roles of glutamates and metal ions in a rationally designed nitric oxide reductase based on myoglobin. *Proc Natl Acad Sci U S A* 107: 8581–8586.

38. Komatsu T, Ohmichi N, Nakagawa A, Zunszain PA, Curry S, Tsuchida E. 2005. O_2 and CO binding properties of artificial hemoproteins formed by complexing iron protoporphyrin IX with human serum albumin mutants. *J Am Chem Soc* 127: 15933–15942.

39. Komatsu T, Nakagawa A, Zunszain PA, Curry S, Tsuchida E. 2007. Genetic engineering of the heme pocket in human serum albumin: Modulation of O2

binding of iron protoporphyrin IX by variation of distal amino acids. *J Am Chem Soc* 129: 11286–11295.

40. Tsuchida E, Sou K, Nakagawa A, Sakai H, Komatsu T, Kobayashi K. 2009. Artificial oxygen carriers, hemoglobin vesicles and albumin-hemes, based on bioconjugate chemistry. *Bioconjug Chem* 20: 1419–1440.

41. Koder RL, Anderson JL, Solomon LA, Reddy KS, Moser CC, Dutton PL. 2009. Design and engineering of an O(2) transport protein. *Nature* 458: 305–309.

42. Fitzgerald MM, Churchill MJ, McRee DE, Goodin DB. 1994. Small molecule binding to an artificially created cavity at the active site of cytochrome c peroxidase. *Biochemistry* 33: 3807–3818.

43. Negron C, Fufezan C, Koder RL. 2009. Geometric constraints for porphyrin binding in helical protein binding sites. *Proteins* 74: 400–416.

44. Suits MD, Jaffer N, Jia Z. 2006. Structure of the *Escherichia coli* O157:H7 heme oxygenase ChuS in complex with heme and enzymatic inactivation by mutation of the heme coordinating residue His-193. *J Biol Chem* 281: 36776–36782.

45. Schneider S, Sharp KH, Barker PD, Paoli M. 2006. An induced fit conformational change underlies the binding mechanism of the heme transport proteobacteria-protein HemS. *J Biol Chem* 281: 32606–32610.

46. Hou S, Larsen RW, Boudko D, Riley CW, Karatan E, Zimmer M, Ordal GW, Alam M. 2000. Myoglobin-like aerotaxis transducers in archaea and bacteria. *Nature* 403: 540–544.

47. Hefti MH, Francoijs KJ, de Vries SC, Dixon R, Vervoort J. 2004. The PAS fold. A redefinition of the PAS domain based upon structural prediction. *Eur J Biochem* 271: 1198–1208.

48. Gilles-Gonzalez MA, Gonzalez G. 2004. Signal transduction by heme-containing PAS-domain proteins. *J Appl Physiol* 96: 774–783.

49. Gilles-Gonzalez MA, Gonzalez G. 2005. Heme-based sensors: Defining characteristics, recent developments, and regulatory hypotheses. *J Inorg Biochem* 99: 1–22.

50. Youn H, Kerby RL, Thorsteinsson MV, Conrad M, Staples CR, Serate J, Beack J, Roberts GP. 2001. The heme pocket afforded by Gly117 is crucial for proper heme ligation and activity of CooA. *J Biol Chem* 276: 41603–41610.

51. Lanzilotta WN, Schuller DJ, Thorsteinsson MV, Kerby RL, Roberts GP, Poulos TL. 2000. Structure of the CO sensing transcription activator CooA. *Nat Struct Biol* 7: 876–880.

52. Leys D, Backers K, Meyer TE, Hagen WR, Cusanovich MA, Van Beeumen JJ. 2000. Crystal structures of an oxygen-binding cytochrome c from Rhodobacter sphaeroides. *J Biol Chem* 275: 16050–16056.

53. Pagola S, Stephens PW, Bohle DS, Kosar AD, Madsen SK. 2000. The structure of malaria pigment beta-haematin *Nature* 404: 307–310.

54. de Villiers KA, Kaschula CH, Egan TJ, Marques HM. 2007. Speciation and structure of ferriprotoporphyrin IX in aqueous solution: Spectroscopic and diffusion measurements demonstrate dimerization, but not mu-oxo dimer formation. *J Biol Inorg Chem* 12: 101–117.

55. Joyce MG, Girvan HM, Munro AW, Leys D. 2004. A single mutation in cytochrome P450 BM3 induces the conformational rearrangement seen upon substrate binding in the wild-type enzyme. *J Biol Chem* 279: 23287–23293.

56. Girvan HM, Marshall KR, Lawson RJ, Leys D, Joyce MG, Clarkson J, Smith WE, Cheesman MR, Munro AW. 2004. Flavocytochrome P450 BM3 mutant A264E undergoes substrate-dependent formation of a novel heme iron ligand set. *J Biol Chem* 279: 23274–23286.

57. Rodrigues ML, Oliveira TF, Pereira IA, Archer M. 2006. X-ray structure of the membrane-bound cytochrome c quinol dehydrogenase NrfH reveals novel haem coordination. *EMBO J* 25: 5951–5960.

58. Rodrigues ML, Scott KA, Sansom MS, Pereira IA, Archer M. 2008. Quinol oxidation by c-type cytochromes: Structural characterization of the menaquinol binding site of NrfHA. *J Mol Biol* 381: 341–350.

59. Adachi S, Nagano S, Ishimori K, Watanabe Y, Morishima I, Egawa T, Kitagawa T, Makino R. 1993. Roles of proximal ligand in heme proteins: Replacement of proximal histidine of human myoglobin with cysteine and tyrosine by site-directed mutagenesis as models for P-450, chloroperoxidase, and catalase. *Biochemistry* 32: 241–252.

60. Kurumbail RG, Stevens AM, Gierse JK, McDonald JJ, Stegeman RA, Pak JY, Gildehaus D, Miyashiro JM, Penning TD, Seibert K, Isakson PC, Stallings WC. 1996. Structural basis for selective inhibition of cyclooxygenase-2 by anti-inflammatory agents. *Nature* 384: 644–648.

61. Luong C, Miller A, Barnett J, Chow J, Ramesha C, Browner MF. 1996. Flexibility of the NSAID binding site in the structure of human cyclooxygenase-2. *Nat Struct Biol* 3: 927–933.

62. Sugimoto H, Oda S, Otsuki T, Hino T, Yoshida T, Shiro Y. 2006. Crystal structure of human indoleamine 2,3-dioxygenase: Catalytic mechanism of O2 incorporation by a heme-containing dioxygenase. *Proc Natl Acad Sci U S A* 103: 2611–2616.

63. Jensen LM, Sanishvili R, Davidson VL, Wilmot CM. 2010. In crystallo post-translational modification within a MauG/pre-methylamine dehydrogenase complex. *Science* 327: 1392–1394.

64. Goff H, Morgan LO. 1976. Amino acid-substituted iron porphyrins. 2. Thermodynamic studies of ligand binding. *Inorganic Chemistry* 15: 2069–2076.

65. Barrick D. 1994. Replacement of the proximal ligand of sperm whale myoglobin with free imidazole in the mutant His-93 → Gly. *Biochemistry* 33: 6546–6554.

66. Barrick D. 1995. Depletion and replacement of protein metal ligands. *Curr Opin Biotechnol* 6: 411–418.

67. Hirst J, Wilcox SK, Ai J, Moenne-Loccoz P, Loehr TM, Goodin DB. 2001. Replacement of the axial histidine ligand with imidazole in cytochrome c peroxidase. 2. Effects on heme coordination and function. *Biochemistry* 40: 1274–1283.

68. Hirst J, Wilcox SK, Williams PA, Blankenship J, McRee DE, Goodin DB. 2001. Replacement of the axial histidine ligand with imidazole in cytochrome c peroxidase. 1. Effects on structure. *Biochemistry* 40: 1265–1273.

69. McRee DE, Jensen GM, Fitzgerald MM, Siegel HA, Goodin DB. 1994. Construction of a bisaquo heme enzyme and binding by exogenous ligands. *Proc Natl Acad Sci U S A* 91: 12847–12851.

70. Barrick D. 2000. Trans-substitution of the proximal hydrogen bond in myoglobin: II. Energetics, functional consequences, and implications for hemoglobin allostery. *Proteins* 39: 291–308.

71. Barrick D, Dahlquist FW. 2000. Trans-substitution of the proximal hydrogen bond in myoglobin: I. Structural consequences of hydrogen bond deletion. *Proteins* 39: 278–290.

72. Zhou WP, Zhong WW, Zhang XH, Ding JP, Zhang ZL, Xia ZW. 2009. Comparison of the crystal structure and function to wild-type and His25Ala mutant human heme oxygenase-1. *Int J Mol Med* 23: 379–387.

73. Roach MP, Ozaki S, Watanabe Y. 2000. Investigations of the myoglobin cavity mutant H93G with unnatural imidazole proximal ligands as a modular peroxide O-O bond cleavage model system. *Biochemistry* 39: 1446–1454.

74. Roach MP, Puspita WJ, Watanabe Y. 2000. Proximal ligand control of heme iron coordination structure and reactivity with hydrogen peroxide: Investigations of the myoglobin cavity mutant H93G with unnatural oxygen donor proximal ligands. *J Inorg Biochem* 81: 173–182.

75. Reddi AR, Reedy CJ, Mui S, Gibney BR. 2007. Thermodynamic investigation into the mechanisms of proton-coupled electron transfer events in heme protein maquettes. *Biochemistry* 46: 291–305.

76. Robinson VL, Smith BB, Arnone A. 2003. A pH-dependent aquomet-to-hemichrome transition in crystalline horse methemoglobin. *Biochemistry* 42: 10113–10125.

77. Caillet-Saguy C, Turano P, Piccioli M, Lukat-Rodgers GS, Czjzek M, Guigliarelli B, Izadi-Pruneyre N, Rodgers KR, Delepierre M, Lecroisey A. 2008. Deciphering the structural role of histidine 83 for heme binding in hemophore HasA. *J Biol Chem* 283: 5960–5970.

78. Wolff N, Deniau C, Letoffe S, Simenel C, Kumar V, Stojiljkovic I, Wandersman C, Delepierre M, Lecroisey A. 2002. Histidine pK(a) shifts and changes of tautomeric states induced by the binding of gallium–protoporphyrin IX in the hemophore HasA(SM). *Protein Sci* 11: 757–765.

79. Ribbing W, Ruterjans H. 1980. Isomeric incorporation of the Haem into monomeric haemoglobins of chironomus thummi thummi. 2. The Bohr effect of the component III explained on a molecular basis and functional differences between the two isomeric structures. *Eur J Biochem* 108: 89–102.

80. Chatake T, Shibayama N, Park SY, Kurihara K, Tamada T, Tanaka I, Niimura N, Kuroki R, Morimoto Y. 2007. Protonation states of buried histidine residues in human deoxyhemoglobin revealed by neutron crystallography. *J Am Chem Soc* 129: 14840–14841.

81. Kovalevsky AY, Chatake T, Shibayama N, Park SY, Ishikawa T, Mustyakimov M, Fisher Z, Langan P, Morimoto Y. 2010. Direct determination of protonation states of histidine residues in a 2 A neutron structure of deoxy-human normal adult hemoglobin and implications for the Bohr effect. *J Mol Biol* 398: 276–291.

82. Mueser TC, Griffith WP, Kovalevsky AY, Guo J, Seaver S, Langan P, Hanson BL. 2010. Hemoglobin redux: Combining neutron and X-ray diffraction with mass spectrometry to analyse the quaternary state of oxidized hemoglobins. *Acta Crystallogr D Biol Crystallogr* 66: 1249–1256.

83. Fufezan C, Zhang J, Gunner MR. 2008. Ligand preference and orientation in b- and c-type heme-binding proteins. *Proteins* 73: 690–704.

84. Zaric SD, Popovic DM, Knapp EW. 2001. Factors determining the orientation of axially coordinated imidazoles in heme proteins. *Biochemistry* 40: 7914–7928.

85. Galstyan AS, Zaric SD, Knapp EW. 2005. Computational studies on imidazole heme conformations. *J Biol Inorg Chem* 10: 343–354.

86. Sarma S, DiGate RJ, Goodin DB, Miller CJ, Guiles RD. 1997. Effect of axial ligand plane reorientation on electronic and electrochemical properties observed in the A67V mutant of rat cytochrome b5. *Biochemistry* 36: 5658–5668.

87. Yi L, Morgan JT, Ragsdale SW. 2010. Identification of a thiol/disulfide redox switch in the human BK channel that controls its affinity for heme and CO. *J Biol Chem* 285: 20117–20127.

88. Allen JW, Barker PD, Daltrop O, Stevens JM, Tomlinson EJ, Sinha N, Sambongi Y, Ferguson SJ. 2005. Why isn't 'standard' heme good enough for c-type and d1-type cytochromes? *Dalton Trans* 3410–3418.

89. Stevens JM, Daltrop O, Allen JW, Ferguson SJ. 2004. C-type cytochrome formation: Chemical and biological enigmas. *Acc Chem Res* 37: 999–1007.

90. Ambler RP. 1991. Sequence variability in bacterial cytochromes c. *Biochim Biophys Acta* 1058: 42–47.

91. Hartshorne RS, Kern M, Meyer B, Clarke TA, Karas M, Richardson DJ, Simon J. 2007. A dedicated haem lyase is required for the maturation of a novel bacterial cytochrome c with unconventional covalent haem binding. *Mol Microbiol* 64: 1049–1060.

92. Hartshorne S, Richardson DJ, Simon J. 2006. Multiple haem lyase genes indicate substrate specificity in cytochrome c biogenesis. *Biochem Soc Trans* 34: 146–149.

93. Aragao D, Frazao C, Sieker L, Sheldrick GM, LeGall J, Carrondo MA. 2003. Structure of dimeric cytochrome c3 from Desulfovibrio gigas at 1.2 A resolution. *Acta Crystallogr D Biol Crystallogr* 59: 644–653.

94. Correia IJ, Paquete CM, Coelho A, Almeida CC, Catarino T, Louro RO, Frazao C, Saraiva LM, Carrondo MA, Turner DL, Xavier AV. 2004. Proton-assisted two-electron transfer in natural variants of tetraheme cytochromes from Desulfomicrobium Sp. *J Biol Chem* 279: 52227–52237.

95. Fulop V, Sam KA, Ferguson SJ, Ginger ML, Allen JW. 2009. Structure of a trypanosomatid mitochondrial cytochrome c with heme attached via only one thioether bond and implications for the substrate recognition requirements of heme lyase. *FEBS J* 276: 2822–2832.

96. Kranz RG, Richard-Fogal C, Taylor JS, Frawley ER. 2009. Cytochrome c biogenesis: Mechanisms for covalent modifications and trafficking of heme and for heme-iron redox control. *Microbiol Mol Biol Rev* 73: 510–528.

97. Kooter IM, Moguilevsky N, Bollen A, van der Veen LA, Otto C, Dekker HL, Wever R. 1999. The sulfonium ion linkage in myeloperoxidase. Direct spectroscopic detection by isotopic labeling and effect of mutation. *J Biol Chem* 274: 26794–26802.

98. Guallar V, Olsen B. 2006. The role of the heme propionates in heme bio-chemistry. *J Inorg Biochem* 100: 755–760.

99. Munakata H, Sun JY, Yoshida K, Nakatani T, Honda E, Hayakawa S, Furuyama K, Hayashi N. 2004. Role of the heme regulatory motif in the heme-mediated inhibition of mitochondrial import of 5-aminolevulinate syn-thase. *J Biochem* 136: 233–238.

100. Dailey TA, Woodruff JH, Dailey HA. 2005. Examination of mitochondrial protein targeting of haem synthetic enzymes: *In vivo* identification of three functional haem-responsive motifs in 5-aminolaevulinate synthase. *Biochem J* 386: 381–386.

101. Goodfellow BJ, Dias JS, Ferreira GC, Henklein P, Wray V, Macedo AL. 2001. The solution structure and heme binding of the presequence of murine 5-aminolevulinate synthase. *FEBS Lett* 505: 325–331.

102. Astner I, Schulze JO, van den Heuvel J, Jahn D, Schubert WD, Heinz DW. 2005. Crystal structure of 5-aminolevulinate synthase, the first enzyme of heme biosynthesis, and its link to XLSA in humans. *EMBO J* 24: 3166–3177.

103. Hunter GA, Ferreira GC. 2009. 5-aminolevulinate synthase: catalysis of the first step of heme biosynthesis. *Cell Mol Biol (Noisy-le-grand)* 55: 102–110.

104. Wu N, Yin L, Hanniman EA, Joshi S, Lazar MA. 2009. Negative feedback maintenance of heme homeostasis by its receptor, Rev-erbalpha. *Genes Dev* 23: 2201–2209.

105. Siegert SW, Holt RJ. 2008. Physicochemical properties, pharmacokinetics, and pharmacodynamics of intravenous hematin: A literature review. *Adv Ther* 25: 842–857.

106. Zunszain PA, Ghuman J, Komatsu T, Tsuchida E, Curry S. 2003. Crystal structural analysis of human serum albumin complexed with hemin and fatty acid. *BMC Struct Biol* 3: 6.

107. Wardell M, Wang Z, Ho JX, Robert J, Ruker F, Ruble J, Carter DC. 2002. The atomic structure of human methemalbumin at 1.9 A. *Biochem Biophys Res Commun* 291: 813–819.

108. Paoli M, Anderson BF, Baker HM, Morgan WT, Smith A, Baker EN. 1999. Crystal structure of hemopexin reveals a novel high-affinity heme site formed between two beta-propeller domains. *Nat Struct Biol* 6: 926–931.

109. Fanali G, De Sanctis G, Gioia M, Coletta M, Ascenzi P, Fasano M. 2009. Reversible two-step unfolding of heme-human serum albumin: A (1)H-NMR relaxometric and circular dichroism study. *J Biol Inorg Chem* 14: 209–217.

110. Wallace AC, Laskowski RA, Thornton JM. 1995. LIGPLOT: A program to gen-erate schematic diagrams of protein-ligand interactions. *Protein Eng* 8: 127–134.

111. Muller-Eberhard U. 1988. Hemopexin. *Methods Enzymol* 163: 536–565.

112. Shipulina N, Smith A, Morgan WT. 2000. Heme binding by hemopexin: Evidence for multiple modes of binding and functional implications. *J Protein Chem* 19: 239–248.

113. Shipulina NV, Smith A, Morgan WT. 2001. Effects of reduction and ligation of heme iron on the thermal stability of heme-hemopexin complexes. *J Protein Chem* 20: 145–154.

114. Baker HM, Anderson BF, Baker EN. 2003. Dealing with iron: Common structural principles in proteins that transport iron and heme. *Proc Natl Acad Sci U S A* 100: 3579–3583.

115. Rosell FI, Mauk MR, Mauk AG. 2005. pH- and metal ion-linked stability of the hemopexin-heme complex. *Biochemistry* 44: 1872–1879.

116. Wolff N, Izadi-Pruneyre N, Couprie J, Habeck M, Linge J, Rieping W, Wandersman C, Nilges M, Delepierre M, Lecroisey A. 2008. Comparative analysis of structural and dynamic properties of the loaded and unloaded hemophore HasA: Functional implications. *J Mol Biol* 376: 517–525.

117. Czjzek M, Letoffe S, Wandersman C, Delepierre M, Lecroisey A, Izadi-Pruneyre N. 2007. The crystal structure of the secreted dimeric form of the hemophore HasA reveals a domain swapping with an exchanged heme ligand. *J Mol Biol* 365: 1176–1186.

118. Brill AS, Williams RJ. 1961. The absorption spectra, magnetic moments and the binding of iron in some haemoproteins. *Biochem J* 78: 246–253.

119. Peisach J, Blumberg WE, Ogawa S, Rachmilewitz EA, Oltzik R. 1971. The effects of protein conformation on the heme symmetry in high spin ferric heme proteins as studied by electron paramagnetic resonance. *J Biol Chem* 246: 3342–3355.

120. Peisach J, Blumberg WE, Wittenberg BA, Wittenberg JB. 1968. The electronic structure of protoheme proteins. 3. Configuration of the heme and its ligands. *J Biol Chem* 243: 1871–1880.

121. Scott RA, Lukehart CM. 2007. *Applications of physical methods to inorganic and bioinorganic chemistry.* Wiley-Interscience.

122. Tezcan FA, Winkler JR, Gray HB. 1998. Effects of ligation and folding on reduction potentials of heme proteins. *J Am Chem Soc* 120: 13383–13388.

123. Mao J, Hauser K, Gunner MR. 2003. How cytochromes with different folds control heme redox potentials. *Biochemistry* 42: 9829–9840.

124. Stellwagen E. 1978. Haem exposure as the determinate of oxidation-reduction potential of haem proteins. *Nature* 275: 73–74.

125. Maes EM, Roberts SA, Weichsel A, Montfort WR. 2005. Ultrahigh resolution structures of nitrophorin 4: Heme distortion in ferrous CO and NO complexes. *Biochemistry* 44: 12690–12699.

126. Olea C, Jr., Kuriyan J, Marletta MA. Modulating heme redox potential through protein-induced porphyrin distortion. *J Am Chem Soc* 132: 12794–12795.

127. Olea C, Boon EM, Pellicena P, Kuriyan J, Marletta MA. 2008. Probing the function of heme distortion in the H-NOX family. *ACS Chem Biol* 3: 703–710.

128. Chen Z, Ost TW, Schelvis JP. 2004. Phe393 mutants of cytochrome P450 BM3 with modified heme redox potentials have altered heme vinyl and propionate conformations. *Biochemistry* 43: 1798–1808.

129. Reid LS, Mauk MR, Mauk AG. 1984. Role of heme propionate groups in cytochrome b5 electron transfer. *J Am Chem Soc* 106: 2182–2185.

130. Das DK, Medhi OK. 1998. The role of heme propionate in controlling the redox potential of heme: Square wave voltammetry of protoporphyrinato IX iron (III) in aqueous surfactant micelles. *J Inorg Biochem* 70: 83–90.

131. Bowman SE, Bren KL. 2010. Variation and analysis of second-sphere interactions and axial histidinate character in c-type cytochromes. *Inorg Chem* 49: 7890–7897.

132. Cho HY, Cho HJ, Kim YM, Oh JI, Kang BS. 2009. Structural insight into the heme-based redox sensing by DosS from *Mycobacterium tuberculosis*. *J Biol Chem* 284: 13057–13067.

133. Badyal SK, Metcalfe CL, Basran J, Efimov I, Moody PC, Raven EL. 2008. Iron oxidation state modulates active site structure in a heme peroxidase. *Biochemistry* 47: 4403–4409.

134. Carlsson GH, Nicholls P, Svistunenko D, Berglund GI, Hajdu J. 2005. Complexes of horseradish peroxidase with formate, acetate, and carbon monoxide. *Biochemistry* 44: 635–642.

135. Hirotsu S, Chu GC, Unno M, Lee DS, Yoshida T, Park SY, Shiro Y, Ikeda-Saito M. 2004. The crystal structures of the ferric and ferrous forms of the heme complex of HmuO, a heme oxygenase of *Corynebacterium diphtheriae*. *J Biol Chem* 279: 11937–11947.

136. Sugishima M, Sakamoto H, Noguchi M, Fukuyama K. 2003. Crystal structures of ferrous and CO-, CN(-)-, and NO-bound forms of rat heme oxygenase-1 (HO-1) in complex with heme: Structural implications for discrimination between CO and O2 in HO-1. *Biochemistry* 42: 9898–9905.

137. Lad L, Wang J, Li H, Friedman J, Bhaskar B, Ortiz de Montellano PR, Poulos TL. 2003. Crystal structures of the ferric, ferrous, and ferrous-NO forms of the Asp140Ala mutant of human heme oxygenase-1: Catalytic implications. *J Mol Biol* 330: 527–538.

138. Geremia S, Garau G, Vaccari L, Sgarra R, Viezzoli MS, Calligaris M, Randaccio L. 2002. Cleavage of the iron-methionine bond in c-type

cytochromes: Crystal structure of oxidized and reduced cytochrome c(2) from *Rhodopseudomonas palustris* and its ammonia complex. *Protein Sci* 11: 6–17.

139. Andreoletti P, Pernoud A, Sainz G, Gouet P, Jouve HM. 2003. Structural studies of Proteus mirabilis catalase in its ground state, oxidized state and in complex with formic acid. *Acta Crystallogr D Biol Crystallogr* 59: 2163–2168.

140. Berghuis AM, Brayer GD. 1992. Oxidation state-dependent conformational changes in cytochrome c. *J Mol Biol* 223: 959–976.

141. Norager S, Legrand P, Pieulle L, Hatchikian C, Roth M. 1999. Crystal structure of the oxidised and reduced acidic cytochrome c3 from Desulfovibrio africanus. *J Mol Biol* 290: 881–902.

142. Ishikawa H, Kato M, Hori H, Ishimori K, Kirisako T, Tokunaga F, Iwai K. 2005. Involvement of heme regulatory motif in heme-mediated ubiquitination and degradation of IRP2. *Mol Cell* 19: 171–181.

143. Dycke C, Bougault C, Gaillard J, Andrieu JP, Pantopoulos K, Moulis JM. 2007. Human iron regulatory protein 2 is easily cleaved in its specific domain: Consequences for the haem binding properties of the protein. *Biochem J* 408: 429–439.

144. Gong W, Hao B, Mansy SS, Gonzalez G, Gilles-Gonzalez MA, Chan MK. 1998. Structure of a biological oxygen sensor: A new mechanism for heme-driven signal transduction. *Proc Natl Acad Sci U S A* 95: 15177–15182.

145. Miyatake H, Mukai M, Park SY, Adachi S, Tamura K, Nakamura H, Nakamura K, Tsuchiya T, Iizuka T, Shiro Y. 2000. Sensory mechanism of oxygen sensor FixL from Rhizobium meliloti: Crystallographic, mutagenesis and resonance Raman spectroscopic studies. *J Mol Biol* 301: 415–431.

146. Key J, Srajer V, Pahl R, Moffat K. 2007. Time-resolved crystallographic studies of the heme domain of the oxygen sensor FixL: Structural dynamics of ligand rebinding and their relation to signal transduction. *Biochemistry* 46: 4706–4715.

147. Ayers RA, Moffat K. 2008. Changes in quaternary structure in the signaling mechanisms of PAS domains. *Biochemistry* 47: 12078–12086.

148. Kurokawa H, Lee DS, Watanabe M, Sagami I, Mikami B, Raman CS, Shimizu T. 2004. A redox-controlled molecular switch revealed by the crystal structure of a bacterial heme PAS sensor. *J Biol Chem* 279: 20186–20193.

149. Park H, Suquet C, Satterlee JD, Kang C. 2004. Insights into signal transduction involving PAS domain oxygen-sensing heme proteins from the X-ray crystal structure of *Escherichia coli* Dos heme domain (Ec DosH). *Biochemistry* 43: 2738–2746.

150. Tanaka A, Shimizu T. 2008. Ligand binding to the Fe(III)-protoporphyrin IX complex of phosphodiesterase from *Escherichia coli* (Ec DOS) markedly enhances catalysis of cyclic di-GMP: Roles of Met95, Arg97, and Phe113 of the putative heme distal side in catalytic regulation and ligand binding. *Biochemistry* 47: 13438–13446.

151. Ito S, Igarashi J, Shimizu T. 2009. The FG loop of a heme-based gas sensor enzyme, Ec DOS, functions in heme binding, autoxidation and catalysis. *J Inorg Biochem* 103: 1380–1385.

152. Ito S, Araki Y, Tanaka A, Igarashi J, Wada T, Shimizu T. 2009. Role of Phe113 at the distal side of the heme domain of an oxygen-sensor (Ec DOS) in the characterization of the heme environment. *J Inorg Biochem* 103: 989–996.

153. Sasakura Y, Yoshimura-Suzuki T, Kurokawa H, Shimizu T. 2006. Structure-function relationships of Ec DOS, a heme-regulated phosphodiesterase from Escherichia coli. *Acc Chem Res* 39: 37–43.

154. Gilles-Gonzalez MA, Caceres AI, Sousa EH, Tomchick DR, Brautigam C, Gonzalez C, Machius M. 2006. A proximal arginine R206 participates in switching of the Bradyrhizobium japonicum FixL oxygen sensor. *J Mol Biol* 360: 80–89.

155. Zhang W, Olson JS, Phillips GN, Jr. 2005. Biophysical and kinetic characterization of HemAT, an aerotaxis receptor from Bacillus subtilis. *Biophys J* 88: 2801–2814.

156. Zhang W, Phillips GN, Jr. 2003. Structure of the oxygen sensor in Bacillus subtilis: Signal transduction of chemotaxis by control of symmetry. *Structure* 11: 1097–1110.

157. Pinakoulaki E, Yoshimura H, Daskalakis V, Yoshioka S, Aono S, Varotsis C. 2006. Two ligand-binding sites in the O2-sensing signal transducer HemAT: Implications for ligand recognition/discrimination and signaling. *Proc Natl Acad Sci U S A* 103: 14796–14801.

158. Roberts GP, Kerby RL, Youn H, Conrad M. 2005. CooA, a paradigm for gas sensing regulatory proteins. *J Inorg Biochem* 99: 280–292.

159. Roberts GP, Thorsteinsson MV, Kerby RL, Lanzilotta WN, Poulos T. 2001. CooA: A heme-containing regulatory protein that serves as a specific sensor of both carbon monoxide and redox state. *Prog Nucleic Acid Res Mol Biol* 67: 35–63.

160. Thorsteinsson MV, Kerby RL, Conrad M, Youn H, Staples CR, Lanzilotta WN, Poulos TJ, Serate J, Roberts GP. 2000. Characterization of variants altered at the N-terminal proline, a novel heme-axial ligand in CooA, the CO-sensing transcriptional activator. *J Biol Chem* 275: 39332–39338.

161. Borjigin M, Li H, Lanz ND, Kerby RL, Roberts GP, Poulos TL. 2007. Structure-based hypothesis on the activation of the CO-sensing transcription factor CooA. *Acta Crystallogr D Biol Crystallogr* 63: 282–287.

162. Komori H, Inagaki S, Yoshioka S, Aono S, Higuchi Y. 2007. Crystal structure of CO-sensing transcription activator CooA bound to exogenous ligand imidazole. *J Mol Biol* 367: 864–871.

163. Zhang T, Rubtsov IV, Nakajima H, Aono S, Yoshihara K. 2006. Effect of mutation on the dissociation and recombination dynamics of CO in transcriptional regulator CooA: A picosecond infrared transient absorption study. *Biochemistry* 45: 9246–9253.

164. Derbyshire ER, Marletta MA. 2009. Biochemistry of soluble guanylate cyclase. *Handb Exp Pharmacol* 17–31.

165. Boon EM, Marletta MA. 2005. Ligand discrimination in soluble guanylate cyclase and the H-NOX family of heme sensor proteins. *Curr Opin Chem Biol* 9: 441–446.

166. Cary SP, Winger JA, Derbyshire ER, Marletta MA. 2006. Nitric oxide signaling: no longer simply on or off. *Trends Biochem Sci* 31: 231–239.

167. Winger JA, Derbyshire ER, Lamers MH, Marletta MA, Kuriyan J. 2008. The crystal structure of the catalytic domain of a eukaryotic guanylate cyclase. *BMC Struct Biol* 8: 42.

168. Ma X, Beuve A, van den Akker F. 2010. Crystal structure of the signaling helix coiled-coil domain of the beta1 subunit of the soluble guanylyl cyclase. *BMC Struct Biol* 10: 2.

169. Nioche P, Berka V, Vipond J, Minton N, Tsai AL, Raman CS. 2004. Femtomolar sensitivity of a NO sensor from clostridium botulinum. *Science* 306: 1550–1553.

170. Pellicena P, Karow DS, Boon EM, Marletta MA, Kuriyan J. 2004. Crystal structure of an oxygen-binding heme domain related to soluble guanylate cyclases. *Proc Natl Acad Sci U S A* 101: 12854–12859.

171. Martin F, Baskaran P, Ma X, Dunten PW, Schaefer M, Stasch JP, Beuve A, van den Akker F. 2010. Structure of cinaciguat (BAY 58–2667) bound to Nostoc H-NOX domain reveals insights into heme-mimetic activation of the soluble guanylyl cyclase. *J Biol Chem* 285: 22651–22657.

172. Russwurm M, Koesling D. 2004. Guanylyl cyclase: NO hits its target. *Biochem Soc Symp* 51–63.

173. Lawson DM, Stevenson CE, Andrew CR, George SJ, Eady RR. 2003. A two-faced molecule offers NO explanation: The proximal binding of nitric oxide to haem. *Biochem Soc Trans* 31: 553–557.

174. Poulos TL. 2006. Soluble guanylate cyclase. *Curr Opin Struct Biol* 16: 736–743.

175. Ma X, Sayed N, Beuve A, van den Akker F. 2007. NO and CO differentially activate soluble guanylyl cyclase via a heme pivot-bend mechanism. *EMBO J* 26: 578–588.

176. Spiro T. 2008. A twist on heme signaling. *ACS Chem Biol* 3: 673–675.

177. Zhang H, Lu M, Zhang Y, Li Z. 2008. Primary response of the sGC heme binding domain to the cleavage of the Fe-His bond. *Bioinformation* 2: 296–300.

178. Capece L, Estrin DA, Marti MA. 2008. Dynamical characterization of the heme NO oxygen binding (HNOX) domain. Insight into soluble guanylate cyclase allosteric transition. *Biochemistry* 47: 9416–9427.

179. Taoka S, Lepore BW, Kabil O, Ojha S, Ringe D, Banerjee R. 2002. Human cystathionine beta-synthase is a heme sensor protein. Evidence that the redox sensor is heme and not the vicinal cysteines in the CXXC motif seen in the crystal structure of the truncated enzyme. *Biochemistry* 41: 10454–10461.

180. Banerjee R, Zou CG. 2005. Redox regulation and reaction mechanism of human cystathionine-beta-synthase: A PLP-dependent hemesensor protein. *Arch Biochem Biophys* 433: 144–156.

181. Singh S, Madzelan P, Stasser J, Weeks CL, Becker D, Spiro TG, Penner-Hahn J, Banerjee R. 2009. Modulation of the heme electronic structure and cystathionine beta-synthase activity by second coordination sphere ligands: The role of heme ligand switching in redox regulation. *J Inorg Biochem* 103: 689–697.

182. Giardina G, Rinaldo S, Johnson KA, Di Matteo A, Brunori M, Cutruzzola F. 2008. NO sensing in *Pseudomonas aeruginosa*: Structure of the transcriptional regulator DNR. *J Mol Biol* 378: 1002–1015.

183. Desai KK, Miller BG. 2010. L-glyceraldehyde 3-phosphate reductase from *Escherichia coli* is a heme binding protein. *Bioorg Chem* 38: 37–41.

184. Hou S, Heinemann SH, Hoshi T. 2009. Modulation of BKCa channel gating by endogenous signaling molecules. *Physiology (Bethesda)* 24: 26–35.

185. Horrigan FT, Heinemann SH, Hoshi T. 2005. Heme regulates allosteric activation of the Slo1 BK channel. *J Gen Physiol* 126: 7–21.

186. Raghuram S, Stayrook KR, Huang P, Rogers PM, Nosie AK, McClure DB, Burris LL, Khorasanizadeh S, Burris TP, Rastinejad F. 2007. Identification of heme as the ligand for the orphan nuclear receptors REV-ERBalpha and REV-ERBbeta. *Nat Struct Mol Biol* 14: 1207–1213.

187. Yin L, Wu N, Curtin JC, Qatanani M, Szwergold NR, Reid RA, Waitt GM, Parks DJ, Pearce KH, Wisely GB, Lazar MA. 2007. Rev-erbalpha, a heme sensor that coordinates metabolic and circadian pathways. *Science* 318: 1786–1789.

188. Rogers PM, Ying L, Burris TP. 2008. Relationship between circadian oscillations of Rev-erbalpha expression and intracellular levels of its ligand, heme. *Biochem Biophys Res Commun* 368: 955–958.

189. Burris TP. 2008. Nuclear hormone receptors for heme: REV-ERBalpha and REV-ERBbeta are ligand-regulated components of the mammalian clock. *Mol Endocrinol* 22: 1509–1520.

190. Kumar N, Solt LA, Wang Y, Rogers PM, Bhattacharyya G, Kamenecka TM, Stayrook KR, Crumbley C, Floyd ZE, Gimble JM, Griffin PR, Burris TP. 2010. Regulation of adipogenesis by natural and synthetic REV-ERB ligands. *Endocrinology* 151: 3015–3025.

191. Woo EJ, Jeong DG, Lim MY, Jun Kim S, Kim KJ, Yoon SM, Park BC, Ryu SE. 2007. Structural insight into the constitutive repression function of the nuclear receptor Rev-erbbeta. *J Mol Biol* 373: 735–744.

192. Phelan CA, Gampe RT, Jr., Lambert MH, Parks DJ, Montana V, Bynum J, Broderick TM, Hu X, Williams SP, Nolte RT, Lazar MA. 2010. Structure of Rev-erbalpha bound to N-CoR reveals a unique mechanism of nuclear receptor-co-repressor interaction. *Nat Struct Mol Biol* 17: 808–814.

193. Marvin KA, Reinking JL, Lee AJ, Pardee K, Krause HM, Burstyn JN. 2009. Nuclear receptors homo sapiens Rev-erbbeta and Drosophila melanogaster E75 are thiolate-ligated heme proteins which undergo redox-mediated ligand switching and bind CO and NO. *Biochemistry* 48: 7056–7071.

Chapter 9

CLINICAL APPLICATIONS OF HEME BIOSYNTHETIC PATHWAY

Photodynamic Therapy with Protoporphyrin IX

Huiying Ding, Baran D. Sumer and Jinming Gao

Overview

Protoporphyrin IX (PpIX) is a naturally occurring porphyrin constituent of heme. In the absence of Fe, PpIX is a potent photosensitizer (PS) that efficiently converts ground state oxygen to singlet oxygen ($^1O_2$) upon photoactivation. PpIX biosynthesis through the heme pathway has already impacted clinical medicine with its implementation into photodynamic therapy for cancer and other diseases. Clinically, 5-aminolevulinic acid (5-ALA), a key intermediate during heme biosynthesis, is administered and results in a rapid and selective accumulation of PpIX in the targeted cells. Irradiation of the targeted cells with the appropriate wavelength of light triggers a series of photochemical reactions, which leads to cellular toxicity. The advantages of photodynamic therapy (PDT) include non-invasive or minimally invasive procedures, high tissue selectivity, quick response and low toxicity. In this chapter, we will review the biosynthesis of PpIX, the mechanism of photosensitization and clinical applications of PDT. In addition to PpIX, other clinically approved PDT agents are also reviewed and discussed in this chapter.

Biosynthesis of PpIX

The detailed biosyntheses of heme and its precursors have been well described in the previous chapters. Here we present a shortened pathway with highlighted key intermediates that are relevant to PDT.

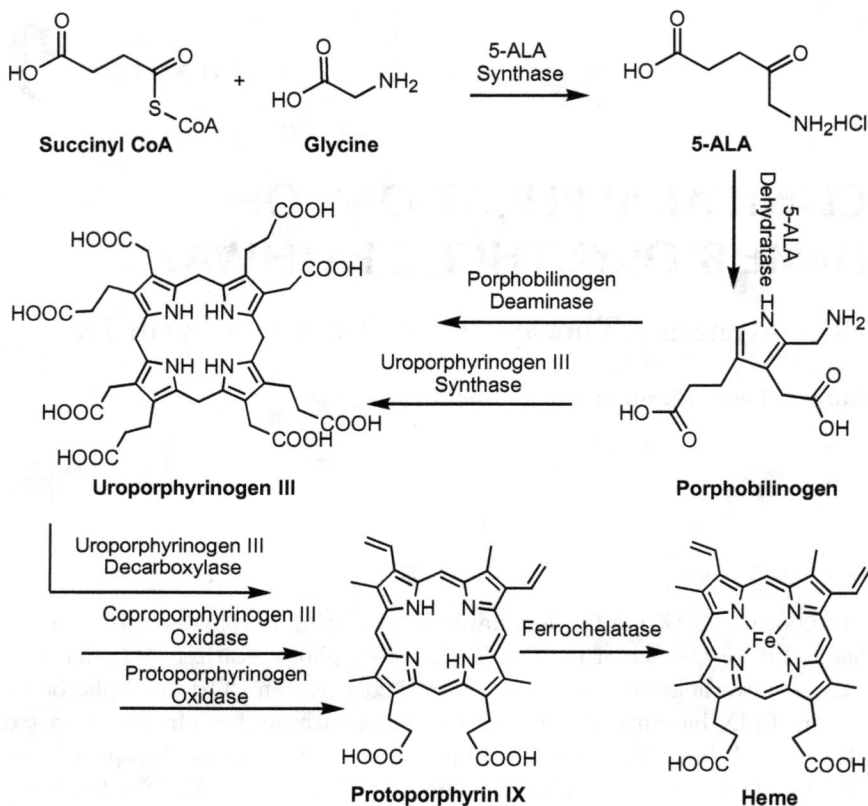

Fig. 1. Biosynthetic pathway of protoporphyrin IX (PpIX) and heme. Mitochondria are involved in three major steps of heme biosynthesis: first, in the formation of 5-ALA from glycine and active succinate; second, in the synthesis of PpIX; third, in the incorporation of Fe^{2+} into the porphyrin ring. The other steps occur in the cytoplasm.

Porphyrins and heme are synthesized in all mammalian cells. PpIX is the direct precursor of heme. Heme biosynthesis consists of a series of enzymatic reactions (Fig. 1), which originate in the mitochondria and continue in the cytoplasm before returning to mitochondria. Deficiency of any one of the enzymes will result in porphyrias, a disease which mostly affects the nervous system or skin. One of the initial enzymatic reactions involves the synthesis of 5-ALA from glycine and succinyl CoA. This reaction is catalyzed by ALA synthetase with pyridoxal phosphate as a cofactor. It is the rate-limiting step in heme biosynthesis (1). Subsequently, ALA dehydratase catalyzes the condensation of two 5-ALA molecules into porphobilinogen, a pyrrole-containing intermediate. After deamination, four porphobilinogen molecules are cyclized to form uroprophyrinogen III, the first macrocyclic intermediate. A series of decarboxylation and oxidation reactions yield protoporphyrin IX inside mitochondria. For heme biosynthesis, the final step is the complexation of Fe^{2+} ions into PpIX by ferrochelatase, a matrix enzyme in the mitochondria. It is important to note that upon Fe chelation, heme loses its ability to generate $^1O_2$ and is no longer photosensitive.

PpIX is synthesized in the mitochondria usually with a higher rate of production in more metabolically active neoplastic cells than in normal tissues. After synthesis, some fractions of PpIX are redistributed from the mitochondria into the perinuclear region of the cytoplasm that contains membrane-rich organelles such as the endoplasmic reticulum, and to the nuclear envelope. Localization into the nuclear membrane could be an explanation for the moderate dark toxicity of PpIX. PpIX can also be found in the plasma membranes and in lysosomes, and thus may be responsible for damage to these subcellular sites following light irradiation (2–4). Due to its very short intracellular half-life, singlet oxygen reacts locally at its sites of production. Therefore, different PS have different sites of action within the cell as a result of different patterns of intracellular distribution (4).

Administration of excess amount of exogenous 5-ALA and the relatively slow conversion of PpIX to heme can result in elevated levels of PpIX. In addition, lack of iron availability for ferrochelatase also plays a role in the accumulation of PpIX. For example, addition of an iron chelator, CP94, greatly increased PpIX accumulation in human skin exposed to 5-ALA (5). PpIX accumulation is more pronounced in rapidly proliferating tumor cells, which not only have reduced ferrochelatase activity, but also enhanced porphobilinogen deaminase activity (6, 7). These factors contribute to the selective increase of PpIX accumulation in cancer cells, which results in improved specificity and therapeutic index of PDT in cancer therapy (8).

Mechanism of PDT

PDT is a relatively new therapeutic modality that has been developed in the last four decades (9). In PDT a photosensitizing drug is administered to the patient and when it is activated by the appropriate wavelength of light in the presence of oxygen, generates $^1O_2$ and reactive oxygen species (ROS) capable of destroying malignant tissues (10, 11). As both the light and the PS can be targeted to the malignant cells, PDT offers the advantage of an effective and selective method of destroying cancer cells without significantly damaging adjacent normal tissues. Two oxidative mechanisms are proposed to contribute to the phototoxicity on target tissues (12) once the PS has been activated to its excited state by light. In the type I mechanism, the PS reacts directly with biological molecules or other substrates to form free radicals that further transfer electrons to molecular oxygen to produce the ROS, superoxide radical anion (O_2^{-}), hydrogen peroxide (H_2O_2) and hydroxyl radical (OH·). In type II mechanism, $^1O_2$ is the main species which is generated via energy transfer from the excited state of PS to molecular oxygen. $^1O_2$ generated during PDT, may have a cytotoxic effect via the initiation of one or more pathways (Fig. 2). The direct cytotoxicity within the tumor microenvironment, involving oxidative damage to nucleic acids, unsaturated lipids, enzymes and cellular membranes, leads to the disintegration of cellular membranes and results in the depletion of intracellular ATP, decrease in intracellular pH, and the induction of apoptosis. In addition to direct damage to tumor

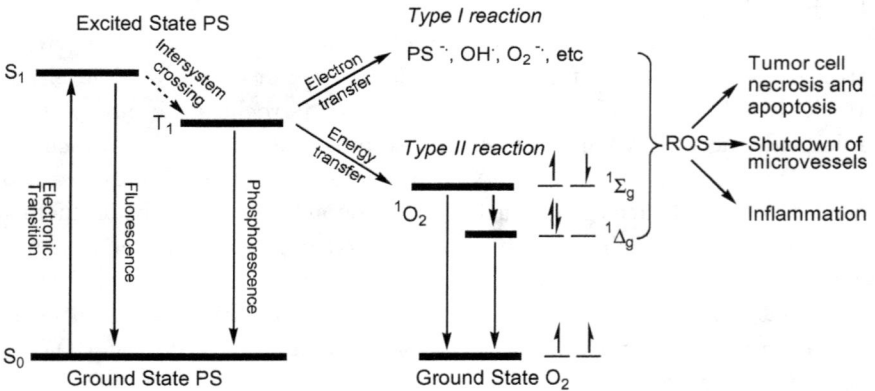

Fig. 2.   Mechanisms of photosensitization through type I and II reactions. In type I mechanism, reactive oxygen species (ROS) are generated. In type II mechanism, singlet oxygen ($^1O_2$) is produced that leads to cellular toxicity. The generated ROS can directly kill tumor cells by the induction of necrosis and/or apoptosis, or can cause destruction of tumor vasculature and produces an acute inflammatory response that attracts leukocytes such as dendritic cells and neutrophils (14).

cells, the effects of $^1O_2$ on the surrounding tissue microcirculation also play a critical role in tumor destruction. Vascular damage can induce tissue hypoxia and lead to cancer cell cytotoxicity by disruption of the blood supply to the tumor. Another important photodynamic action is mediated by the inflammatory response (13). The loss of vascular homeostasis followed by massive damage to vascular and perivascular regions in PDT-treated tumors leads to critical initiating events including the activation of complement, and tumor antigen specific immune responses as well as activation of poly(ADP-ribose)polymerase (PARP).

Photosensitizers (PS)

PS is a critical element in PDT. There are two approaches to the administration of photosensitizers. One is exogenous and involves administration of the photosensitizer intravenously and relies on selective uptake of the drug by the target tissue such as a tumor. The second is to rely on endogenous photosensitization whereby the target cells convert a precursor into the photoreactive compound. An ideal photosensitizer should have the following characteristics: (1) chemical purity and known composition, (2) toxicity only in the presence of light, (3) preferential retention in the target tissue, (4) rapid excretion following treatment, (5) low systemic toxicity, (6) high quantum yield for the photochemical process (high triplet yields, Φ_T, and long triplet lifetimes, τ_T, to generate singlet oxygen and other reactive oxygen species), and (7) strong absorbance with a high extinction coefficient ε, in the 630–800 nm range where tissue penetration of light is at a maximum while still being energetic enough to produce singlet oxygen. To date, only two precursors and three photosensitizers have been approved by the FDA in the US, which include levulan®(5-ALA), Metvixia® (methyl) 5-ALA, Photofrin® (profimer sodium), and Visudyne® (verteporfin). Figure 3 shows the molecular structures for Photofrin® and Visudyne®.

Levulan® (5-ALA, 5-Aminolevulinic acid hydrochloride)

As a key intermediate for heme biosynthesis, PpIX provides the most biocompatible choice for PDT. PpIX has a strong UV-Vis absorption in the 380–450 nm region. Its spectroscopy consists of a Soret band with the maximum absorption peak at 404 nm (15), and Q-bands at approximately 510, 545, 580 and 630 nm. Although the Q-bands are 10–20-folds smaller than the peak absorption in the Soret band, most clinical studies use

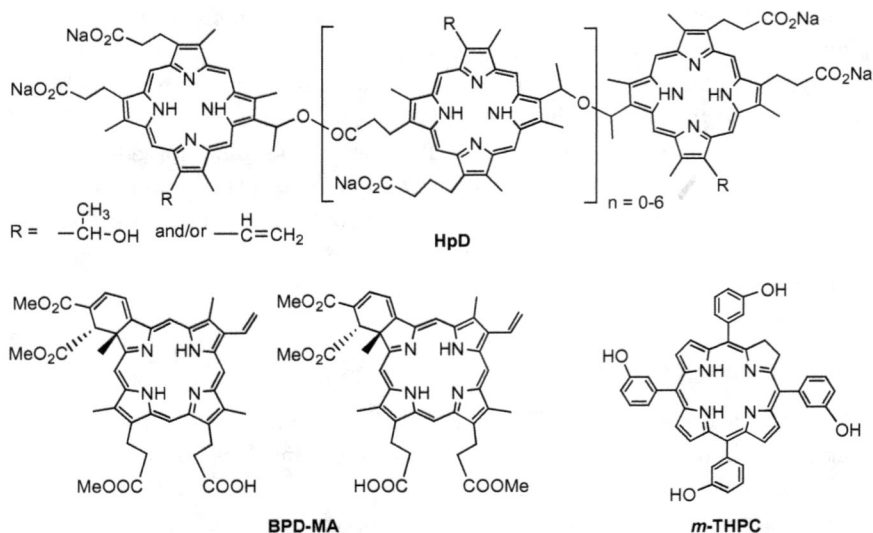

Fig. 3. Molecular structures of Photofrin® (HpD), Visudyne® (BPD-MA) and Foscan® (*m*-THPC).

625–633 nm light to achieve deeper tissue penetration (16). However, direct usage of PpIX as a PDT drug is limited because of several factors. First of all, PpIX is hydrophobic and has poor water solubility, which makes it difficult to deliver a sufficient dose to the malignant tissues or cells. Secondly, PpIX aggregates even at very low concentrations (less than 1 μM), which significantly decreases the efficiency for singlet oxygen generation and phototoxicity.

Since 5-ALA is a key intermediate and its synthesis is a rate limiting step during PpIX and heme biosyntheses, exogenous administration of 5-ALA and its derivatives becomes a key strategy to achieve photodynamic therapy. 5-ALA-hydrochloride is a white powder that dissolves easily in water, giving an acidic solution (pH = 2.2 at 1%) with absorption bands in the ultraviolet spectral range. It is not stable in aqueous solution in the neutral to basic pH range. A major drawback to 5-ALA based PDT is its poor diffusion through biological membranes because of its low lipophilicity (17). As a result, high doses of 5-ALA must be administered in order to reach clinically relevant levels of PpIX. One technique to mitigate this problem involves the use of esterified 5-ALA. It has been shown that 30–150-folds lower concentrations of hexyl, heptyl or octyl 5-ALA-esters can result in the same amount of PpIX in carcinoma cells compared to non-esterified ALA (18). Hexyl 5-ALA ester has recently been approved for the detection of bladder cancer in Europe. In the US, FDA has approved its Metvixia® cream for the treatment of actinic keratosis, a precancerous skin lesion.

The biosynthesis of PpIX after administration of 5-ALA has been used in the treatment of premalignant and malignant lesions, such as actinic keratosis, brain cancer, and cutaneous T-cell lymphoma. The concept of exogenous topical administration of 5-ALA leading to the endogenous cellular production of photosensitizing concentrations of PpIX, was first established by Kennedy *et al.* (19) for the treatment of skin cancers and their precursors. The fluorescence and photosensitizing properties of PpIX were used for the detection and photodestruction of malignant cells.

5-ALA is frequently applied topically or systemically for PDT of skin tumors (such as basal cell carcinomas [BCC] and squamous cell carcinomas [SCC]) as well as in the diagnostic evaluation of tumors of the skin, bladder, gastrointestinal tract and lung. 5-ALA is hydrophilic and does not easily penetrate through intact skin nor through cell membranes. When 5-ALA is topically administrated and allowed to penetrate into the skin for several hours, subsequent exposure to light of the appropriate wavelength can cause the selective apoptosis of actinic keratosis (AK). Dijkstra *et al.* reported the successful treatment of actinic keratosis, Bowen's disease and superficial basal cell carcinoma using violet light (400–450 nm) and topical 5-ALA (20). This topical system is licensed for use with the Blu-U® Blur Light photodynamic therapy illuminator, a specialized illumination source designed to provide a uniform distribution of blue light to the target areas. After treatment with topical Levulan Kerastick plus the Blu-U® Blur Light, 88% of patients had 75% or more of AK lesions cleared. In another study, 14 patients were treated with 635-nm light and topical 5-ALA, with a complete response seen in 100% of cutaneous lymphoma lesions and 42% of the BCCs treated, with only 3 of 19 lesions failing to respond (21). Systemic 5-ALA was used in the treatment of tumors in 10 patients with various cancers (5 duodenal, 3 esophageal, 2 colorectal). Six hours following oral 5-ALA administration, a laser emitting light at 628 nm was used to illuminate the tumors. In 8 of 10 tumors, endoscopic and histological necrosis was induced 6 days after treatment (22).

5-ALA induced PpIX is not completely specific for malignant tissue, and there is PpIX-induced photosensitization of normal tissues. This effect is mainly caused by the relatively slow conversion of PpIX to heme and places normal tissues at risk for photosensitivity. However, such damage can be minimized by relying on bleaching of the photosensitizer. 5-ALA–induced PpIX is rapidly photobleached in normal cells when irradiated with light;

that is, PpIX is destroyed by auto-oxidation by reacting with ROS. Thus, a low concentration of tissue PpIX can be photobleached before the photodynamic threshold for tissue damage occurs. This phenomenon makes it possible to "overdose" the treatment field to get maximum PDT effect without causing serious damage to normal tissue. However, malignant cells will only be destroyed if sufficient PpIX accumulates so that there is a loss of viability before photobleaching can reduce the PpIX concentration to a non-toxic level (2).

Photofrin® (Hematoporphyrin Derivative, HpD)

Photofrin® was the first PS that received FDA approval for clinical use. HpD is actually a mixture of monomers, dimers, and oligomers of hematoporphyrin derivatives. The complex mixture is required for clinical activity. Similar combinations of hematoprophyrin-based derivatives are also available from different companies in different parts of the world, such as Photogem® and Photosan® (23). The clinical activity of these various formulations are not interchangeable. In the US, Photofrin® is approved for early and late endobronchial lesions as well as for Barrett's esophagus and for palliation of obstructing esophageal cancers. The drug is approved worldwide for a number of additional uses such as the treatment of bladder cancer. It does, however, have several disadvantages including a slow rate of tissue clearance leading to longer systemic photosensitivity where patients have to avoid light exposure, a relatively low level of tissue specificity and the need for intravenous administration.

Photofrin® has a maximum absorption at 395 nm at high concentration. An absorption band appears with $\lambda_{max} = 400$ nm at low concentration. It is proposed that at high concentrations, a redistribution takes place between the aggregate form and dimers, and at low concentrations, the band at $\lambda_{max} = 400$ nm corresponds to the monomer (24). The emission spectrum has a higher band peaked at about 617 nm and a lower one peaked at about 677 nm. The value of singlet oxygen quantum yield was 0.64 for HpD in methanol (25, 26).

Visudyne® (Verteporfin, benzoporhyrin derivative, BPD-MA)

Visudyne® was approved by the FDA to treat tumor- and age-related macular degeneration in 2000. It is a benzoporphyrin derivative (BPD-MA), which is clinically active when formulated with liposomes (27). As shown in Fig. 3, it is a 1:1 mixture of two regioisomers with a maximal absorption peak at 688 nm. The advantage of Visudyne® is that it can be active with longer wavelengths of light allowing for deeper penetration of the light into tissue. The drug is rapidly accumulated and cleared so that skin photosensitization is minimal. In general, treatment occurs 15–30 mins after injection. Most of the clinical response from Visudyne® sensitization is based on vascular disruption. Therefore, this drug is ideal for lesions dependent on neovasculature. Visudyne® is transported in the plasma primarily by lipoproteins, and exhibits a bi-exponential elimination pharmacokinetics with a terminal elimination half-life of approximately 5–6 hours. Co-localization studies of BPD-MA and Rhodamine 123 showed that it largely accumulates in mitochondria, but also can be found to a lesser extent in the cytoplasm (28).

BPD-MA shows a strong absorbance in the red light portions of the visible spectra (λ_{max} = 688 nm, ε = 33 000 M^{-1} cm^{-1} in organic solvents). The fluorescence peak of BPD-MA in PBS is at 710 nm. BPD-MA presents rather high singlet oxygen quantum yield (Φ_Δ = 0.78) (29, 30).

Foscan® (temoporfin, meta-tetrahydroxyphenylchlorin, or m-THPC)

Compared to porphyrins, chlorins offer increased absorption in the farthest-red band, thus enabling the use of a light with deeper penetration in biological tissues. Moreover, m-THPC presents good singlet oxygen quantum yields, making it an overall better photosensitizer in the clinical setting. m-THPC is a second-generation photosensitizer clinically used for the treatment of human mesothelioma and for gynecological, respiratory, and head and neck cancers, and has been approved for clinical use in Europe (31). m-THPC has a characteristic absorption spectrum consistent with other chlorin-type compounds with two main peaks centered at 417 nm (Soret band) and 650 nm (Q band). The typical fluorescence emission peak is centered at 654 nm. The aggregated species are characterized by a decrease of extinction in all bands combined with bathochromic shifts. The bathochromic shift of the Soret band to 427 nm is the most pronounced. These features are consistent with the formation of J-type aggregates. The triplet and $^1O_2$ yields of m-THPC reported for m-THPC were 0.89 and 0.6, respectively (32). m-THPC has shown better therapeutic gains with longer excitation wavelength at 652 nm (33).

Table 1.   The photophysical properties of PpIX, HpD, BPD-MA, and *m*-THPC.

PS	λ_{abs}^{max}(nm) (ε)[a]	λ_{em}^{max} (nm)	Φ_f (τ_f)	Φ_Δ
PpIX [34]	408 (17,000)[b]	618 (H_2O)	0.155 (16 ns)	0.56
	506 (16,400)	633 (MeOH)		
	542 (13,300)			
	577 (8,600)			
	630 (6,200)			
HpD [35, 36] (mixtures)	400	617	—	0.64
BPD-MA [30]	688 (33,000)	710 (DPBS)	0.05 (5.2 ns)	0.78
m-THPC [37]	417 (81,000)	654 (EtOH)	0.24 (14 ns)	0.60 [32]
	517 (10,800)			
	542 (6,800)			
	593 (4,300)			
	650 (21,800)			

The fluorescence quantum yield Φ_f, fluorescence lifetime τ_f, and singlet oxygen quantum yield Φ_Δ are compared for the four approved photosensitizers.
[a] Molar extinction coefficients are expressed in $M^{-1}cm^{-1}$; [b] in DMSO.

Summary

Clinical application of 5-aminolevulinic acid (5-ALA) and its esters provides a great example of translating the knowledge in the heme biosynthetic pathway into the photodynamic therapy of cancer and other diseases. Admini-stration of 5-ALA, a key intermediate during heme biosynthesis, leads to a selective accumulation of PpIX in the targeted malignant cells. Light irradiation of the targeted cells triggers a series of photochemical reactions, which results in tissue-specific toxicity. In addition to PpIX, other types of photosensitizers are actively developed to increase the PDT efficacy with lower toxicity. PDT is a field at the interface of chemistry, physics, biology and medicine. Fundamental understanding of the molecular and cellular mechanisms of PDT actions will be critical to further develop new and improved photosensitizers for clinical use.

Acknowledgement

This work is supported by NIH/NCRR (UL1 RR024982-02 (Packer)).

References

1. Batlle AM. 1993. Porphyrins, porphyrias, cancer and photodynamic therapy — a model for carcinogenesis. *J Photochem Photobiol B* 20: 5–22.
2. Webber J, Kessel D, Fromm D. 1997. Photodynamic therapy using endogenous photosensitization for gastrointestinal tumors. *Yale J Biol Med* 70: 127–137.
3. Peng Q, Moan J, Warloe T, Nesland JM, Rimington C. 1992. Distribution and photosensitizing efficiency of porphyrins induced by application of exogenous 5-aminolevulinic acid in mice bearing mammary carcinoma *Int J Cancer* 52: 433–443.
4. Moan J, Berg K. 1992. Photochemotherapy of cancer: Experimental research, *Photochem Photobiol* 55: 931–948.
5. Casas A, Batlle AM, Butler AR, Robertson D, Brown EH, MacRobert A, Riley PA. 1999. Comparative effect of ALA derivatives on protoporphyrin IX production in human and rat skin organ cultures. *Br J Cancer* 80: 1525–1532.
6. Kennedy JC, Pottier RH. 1992. Endogenous protoporphyrin IX, a clinically useful photosensitizer for photodynamic therapy. *J Photochem Photobiol B* 14: 275–292.
7. Kalka K, Merk H, Mukhtar H. 2000. Photodynamic therapy in dermatology, *J Am Acad Dermatol* 42: 389–413; quiz 414–386.
8. McCarron PA, Donnelly RF, Andrews GP, Woolfson AD. 2005. Stability of 5-aminolevulinic acid in novel non-aqueous gel and patch-type systems intended for topical application. *J Pharm Sci* 94: 1756–1771.
9. Huang Z, Xu H, Meyers AD, Musani A⁻, Wang L, Tagg R, Barqawi AB, Chen YK. 2008. Photodynamic therapy for treatment of solid tumors-potential and technical challenges. *Technol Cancer Res Treat* 7: 309–320.
10. Grossweiner LI, Patel AS, Grossweiner JB. 1982. Type I and type II mechanisms in the photosensitized lysis of phosphatidylcholine liposomes by hematoporphyrin. *Photochem Photobiol* 36: 159–167.
11. Fuchs J, Thiele J. 1998. The role of oxygen in cutaneous photodynamic therapy. *Free Radic Biol Med* 24: 835–847.
12. Dougherty TJ, 1987. Photosensitizers: Therapy and detection of malignant tumors. *Photochem Photobiol* 45: 879–889.
13. Paras NP, 2003. *Introduction of Biophotonics.* John Wiley & Sons Ltd.
14. Gollnick SO, Owczarczak B, Maier P. 2006. Photodynamic therapy and anti-tumor immunity. *Lasers Surg Med* 38: 509–515.

15. Perez JLJ, Cruz-Orea A, Ramon-Gallegos E, Fuentes RG, Ramirez JFS, 2008. Photoacoustic Spectroscopy to determine *in vitro* the non radiative relaxation time of protoporphyrin IX solution containing gold metallic nanoparticles. *Eur Phys J-Spec Top* 153: 353–356.

16. Eichler J, Knof J, Lenz H. 1977. Measurements on Depth of Penetration of Light (0.35–1.0 μm) in Tissue. *Radiat Environ Bioph* 14: 239–242.

17. Uehlinger P, Zellweger M, Wagnieres G, Juillerat-Jeanneret L, van den Bergh H, Lange N. 2000. 5-Aminolevulinic acid and its derivatives: Physical chemical properties and protoporphyrin IX formation in cultured cells. *J Photochem Photobiol B* 54: 72–80.

18. Gaullier JM, Berg K, Peng Q, Anholt H, Selbo PK, Ma LW, Moan J. 1997. Use of 5-aminolevulinic acid esters to improve photodynamic therapy on cells in culture. *Cancer Research* 57: 1481–1486.

19. Kennedy JC, Pottier RH, Pross DC. 1990. Photodynamic therapy with endogenous protoporphyrin IX: Basic principles and present clinical experience. *J Photochem Photobiol B* 6: 143–148.

20. Dijkstra AT, Majoie IM, van Dongen JW, van Weelden H, van Vloten WA. 2001. Photodynamic therapy with violet light and topical 6-aminolaevulinic acid in the treatment of actinic keratosis, Bowen's disease and basal cell carcinoma. *J Eur Acad Dermatol Venereol* 15: 550–554.

21. Wang I, Bauer B, Andersson-Engels S, Svanberg S, Svanberg K, 1999. Photodynamic therapy utilising topical delta-aminolevulinic acid in non-melanoma skin malignancies of the eyelid and the periocular skin. *Acta Ophthalmol Scan* 77: 182–188.

22. Kelty CJ, Brown NJ, Reed MWR, Ackroyd R, 2002. The use of 5-aminolaevulinic acid as a photosensitiser in photodynamic therapy and photodiagnosis. *Photoch Photobio Sci* 1: 158–168.

23. Dougherty TJ. 1993. Photodynamic therapy. *Photochem Photobiol* 58: 895–900.

24. Potomskene YV, Rotomskis RI. 1988. Studies of temperature dependence of disaggregation process and appearance of a new spectral form of hematoporphyrin diacetate. *J Appl Spectrosc* 49: 621–624.

25. Wolbarsht ML. 1991. *Laser Applications in Medicine and Biology*. Springer.

26. Tanielian C, Schweitzer C, Mechin R, Wolff C. 2001. Quantum yield of singlet oxygen production by monomeric and aggregated forms of hematoporphyrin derivative. *Free Radic Biol Med* 30: 208–212.

27. Houle JM, Strong A. 2002. Clinical pharmacokinetics of verteporfin. *J Clin Pharmacol* 42: 547–557.

28. Runnels JM, Chen N, Ortel B, Kato D, Hasan T. 1999. BPD-MA-mediated photosensitization *in vitro* and *in vivo*: Cellular adhesion and beta1 integrin expression in ovarian cancer cells. *Br J Cancer* 80: 946–953.

29. Gillies R, Kollias N, Hasan T, Diddens H. 1996. Spectral characterization of the benzoporphyrin derivative monoacid ring-A photoproduct formed in fetal calf solutions during irradiation with 694 nm continuous-wave radiation. *J Photochem Photobiol B* 33: 87–90.

30. Aveline B, Hasan T, Redmond RW. 1994. Photophysical and photosensitizing properties of benzoporphyrin derivative monoacid ring A (BPD-MA). *Photochem Photobiol* 59: 328–335.

31. Atif M, Dyer PE, Paget TA, Snelling HV, Stringer MR. 2007. Two-photon excitation studies of m-THPC photosensitizer and photodynamic activity in an epithelial cell line. *Photodiagnosis Photodyn Ther* 4: 106–111.

32. Redmond RW, Gamlin JN. 1999. A compilation of singlet oxygen yields from biologically relevant molecules. *Photochem Photobiol* 70: 391–475.

33. Ris HB, Krueger T, Giger A, Lim CK, Stewart JC, Althaus U, Altermatt HJ. 1999. Photodynamic therapy with mTHPC and polyethylene glycol-derived mTHPC: A comparative study on human tumour xenografts. *Br J Cancer* 79: 1061-1066.

34. Scolaro LM, Castriciano M, Romeo A, Patane S, Cefali E, Allegrini M. 2002. Aggregation behavior of protoporphyrin IX in aqueous solutions: Clear evidence of vesicle formation. *J Phys Chem B* 106: 2453–2459.

35. Tanimoto Y, Takayama M, Inokuchi I, Itoh M, Miyoshi N, Hisazumi H. 1986. Ground and excited states of hematoporphyrin and its derivatives. *Bull Chem Soc Jpn* 59: 3308–3310.

36. Land EJ, 1984. Porphyrin Phototherapy of Human Cancer. *Int J Radiat Biol* 46: 219–223.

37. Gravier J, Schneider R, Frochot C, Bastogne T, Schmitt F, Didelon J, Guillemin F, Barberi-Heyob M. 2008. Improvement of meta-tetra(hydroxyphenyl)chlorin-like photosensitizer selectivity with folate-based targeted delivery. Synthesis and *in vivo* delivery studies. *J Med Chem* 51: 3867–3877.

INDEX

www.ingramcontent.com/pod-product-compliance
Lightning Source LLC
Chambersburg PA
CBHW050559190326
41458CB00007B/2103